THE FIELD DESCRIPTION OF

Engineering Soils and Rocks

The Geological Society of London Handbook Series
published in association with
the Open University Press comprises

Barnes: *Basic Geological Mapping* (Second edition)
Fry: *The Field Description of Metamorphic Rocks*
McClay: *The Mapping of Geological Structures*
Milsom: *Field Guide to Geophysics*
Thorpe and Brown: *The Field Description of Igneous Rocks*
Tucker: *The Field Description of Sedimentary Rocks*

Professional Handbooks in the Series

Brassington: *Field Hydrogeology*
Clark: *The Field Guide to Water Wells and Boreholes*

HANDBOOK SERIES EDITOR – K. G. COX

THE FIELD DESCRIPTION OF

Engineering Soils and Rocks

Graham West

Transport and Road Research Laboratory

OPEN UNIVERSITY PRESS
MILTON KEYNES • PHILADELPHIA

Open University Press
Celtic Court
22 Ballmoor
Buckingham
MK18 1XW

and
1900 Frost Road, Suite 101
Bristol, PA 19007, USA

First published 1991

British Library Cataloguing in Publication Data

West, Graham
The field description of engineering soils and rocks.
1. Engineering geology
I. Title II. series
624.151

ISBN 0-335-15208-2

Library of Congress Cataloging in Publication Data available

Typeset by Scarborough Typesetting Services
Printed in Great Britain by Biddles Limited,
Guildford and King's Lynn

For Charlotte and Melissa

Contents

Preface

This book is a *basic field guide* to the description of engineering soils and rocks. As well as telling you how to describe soils and rocks it also describes some of the situations in which they have to be logged on engineering sites. It is the hope of the author that the book will not be seen on the engineering geologist's bookshelf; its place is in the pocket on site and it is hoped that the book will eventually disintegrate with use in the field.

Engineering geologists face possible dangers when working in the field – from heavy equipment, collapsing trenches, and unstable tunnels and shafts – and the reader is reminded of the paramount importance of following up on the references to technical and statutory literature on safety procedures given in this book. This is a *pocket guide book* and not a definitive instruction manual.

The author is grateful to colleagues who helped in the preparation of the book. Those that kindly supplied illustrations are acknowledged at the appropriate places in the book, but thanks are also due to Martin Culshaw and Alan Forster of the British Geological Survey for helpful discussions. John Perry is thanked for reading the whole book in draft form and making useful comments.

The author also thanks his employer, the Transport and Road Research Laboratory for permission to write the book and for assistance with its preparation. In particular Geoff Margason, the former Director of the Laboratory is thanked for permission to publish, and Dr Myles O'Reilly, the Head of the Ground Engineering Division is thanked for encouragement. However, the views expressed in the book are those of the author and not those of the Laboratory nor of the Department of Transport of which it forms a part.

Dr Keith Cox, the *Handbook Series* Editor, and Richard Baggaley, Commissioning Editor of the Open University Press, are thanked for inviting the author to write the book and smoothing the way to its publication.

The aim of the *Handbook Series* is to provide concise, authoritative, practical guides to field geology: I hope this book will meet these needs for the beginner engineering geologist, and to the extent that this will improve site investigation practice, the promoters of civil engineering works will also benefit.

1 Introduction

The objective of this book is to show how to describe soils and rocks for engineering purposes. It is intended to be used in the field or in the site laboratory and to represent recent practice. A general knowledge of geology is assumed as is some familiarity with civil engineering needs. In particular, the reader's attention is drawn to the other *Handbooks* in this series that are listed at the front of this book, and the advice given in them will not be repeated here; they will be referred to in this book where appropriate. The book is aimed at engineering geology undergraduate and postgraduate students, and civil engineering undergraduates who are specializing in geotechnics; it may also be useful to qualified engineering geologists and civil engineers who are embarking on field geotechnical work for the first time.

1.1 Scope

By *soil* and *rock* in this book we mean what civil engineers understand by these terms. Thus soil is any naturally occurring loose or soft material resulting from the weathering or breakdown of rock or the decay of vegetation. It includes gravels, sands, silts, clays and peats. Rock is any hard, indurated or consolidated massive geological material. Note that these definitions do not necessarily correspond with geologists' categories of Drift and Solid (e.g. London Clay is classified by the engineer as a soil although it is a Solid geological formation) nor does this meaning of soil correspond with that of the geologist or pedologist. Civil engineers' definitions of soil and rock can be very loosely thought of as soils being materials that can be moved by bulldozer or scraper whilst rocks are materials that require blasting or ripping.

Carrying out a field description of engineering soils and rocks requires both observations and tests, and the book gives advice on how to do them. It limits itself to work that can be done by one or two people in the field or in a small site laboratory. Soil and rock testing in a main laboratory, or site measurements and testing that requires a geotechnical team or elaborate instrumentation is outside the scope of the book, as are the reasons for wanting to describe soils and rocks and the analysis and reporting of results after this has been done.

1.2 Approach

First of all a description is given of some simple equipment that you will need and some general advice on using it in the field. Specialist pieces of equipment, however, are dealt with in the later chapters along with their particular application. Then follows a chapter on maps, plans and air photographs because these are of such basic importance to working in the field or on site where the soil and rock description will be made, and because of the need to record the locations of your observations and tests. Chapters 4 and 5 deal with the description of soils and rocks, a careful distinction being made between mass properties and material properties. Chapter 6 describes the measurement of rock and soil strength. The book continues with advice on describing soils and rocks in natural and artificial exposures, and during the course of soft ground boring. Finally, the book concludes with advice on logging soils and rocks in excavations.

In all the work described, although the value of good qualitative soil and rock descriptions is recognized, emphasis is placed on a quantitative approach wherever possible. Certain parts of the book are based on the recommendations of *British Standard Code of Practice for Site Investigations (BS 5930:1981)* because this Code is the one that most British engineering geologists will be working to. However, some adaptation and simplification has been made to suit the needs of a field handbook, and some rationalization has been made to the system of rock description. If a point is at issue, the Code should be referred to and taken as the authority. The extracts from *BS 5930:1981* are reproduced by kind permission of the British Standards Institution; complete copies can be obtained from the address given at the end of the Bibliography. Where appropriate, reference is also made to United States standards and practice; again useful addresses are given at the end of the book.

Portable electronic calculators with memory stores and data processing facilities will be possessed by many readers, to whom it will occur that in several of the operations described in this *Handbook* (e.g. making a series of Schmidt hammer tests) the values could be entered directly into the calculator as each observation is made, and then processed to give the desired output – say mean and standard deviation. This temptation should be resisted and the primary data should always be recorded in notebook or logsheet. Apart from the danger of 'finger trouble', one reason for this is that unless the primary data are recorded they cannot be used for any subsequent analysis that is not anticipated at the time the data are collected. Of course this is not to say that calculators should not be used to process and analyse data on site.

In many parts of the *Handbook* the value of supplementing careful observation and description of soils and rocks by means of photographs is stressed, and advice is given on some simple photographic methods. A single-lens-reflex camera with interchangeable lenses is recommended. This, together with much of the equipment described in the book, will be of very small cost compared with the value of the engineering works you will be involved with, so get items of the best quality where there is a choice. However a word of caution is necessary here – don't substitute the taking of photographs for proper soil and rock description and logging. Similarly, you are urged to resolve all uncertainties while still on site or in the field and not to try and do this back in the office afterwards.

1.3 Limitations

This *Handbook* is a pocket guide and not a definitive instruction manual, and this means there are some limitations that must be understood. In particular, engineering geologists face a number of potential dangers when working in the field or on civil engineering sites, and the reader's attention is drawn to the necessity of referring to the specific standards, codes of practice and statutory regulations cited in this book at the appropriate places in the following chapters.

In the same way, the methods of describing soils and rocks in the field given in the *Handbook* are the essential features of procedures described in more detail in the references cited, and it is recommended that the beginner engineering geologist should consult the originals before starting field work.

The descriptions of how to use instruments, apparatus and techniques are also concise summaries, but it is hoped they contain the essential information needed for working in the field. Before using any piece of equipment, study the supplier's instructions, and remember that different models of the same apparatus may differ in detail arrangement. When using any instrument, ask youself if it is giving a reasonable result, and devise and carry out your own checks on its performance and accuracy.

Finally, although the advice given in the *Handbook* is given in good faith, it is emphasized that the reader follows it on his or her own responsibility.

2
Equipment

The basic items of field equipment and instruments you will need and their use have been well described in Barnes's *Basic Geological Mapping* in the *Handbook* series and this advice will not be repeated here except to provide a checklist of items and to comment on the special needs of the engineering geologist.

The following equipment comprise the basic kit:

Notebook, pencil, measuring tape, arrows, compass-clinometer, camera, flash gun, field glasses. Topographical and geological maps, site plans, scales, protractor, map case. Air photographs, viewing board, viewing aid, pocket stereoscope, 'Chinagraph' pencil. Hand auger, geological hammer, chisel, trowel, polythene bags, ties and labels. Penknife, hand lens (×10), 50 per cent hydrochloric acid (in small polythene dropping bottle with screw cap), feeler gauge.

Some of these items are illustrated in Fig. 2.1. The rainproof survey book shown makes a convenient field notebook. The use of the penknife, hand lens and acid bottle is exemplified in Table 2.1. The three common white or colourless minerals (quartz, calcite and gypsum) may be distinguished by the hardness test: gypsum if present in soils or rocks may have a deleterious effect on concrete. The use of the air photograph items of equipment is described in Section 3.4. Additional special items for engineering geological work will now be described.

2.1 Boring and prospecting tool

For investigating a site at the preliminary stage before mechanized boring equipment is available, the portable hand-operated 'Mackintosh' boring and prospecting tool shown in Fig. 2.2, will be useful. Using the auger boring tool and the water-sampling tube, small samples of soil and ground water can be obtained down to 15-m depth, and using the driving point and hammer, soft alluvial soils and peat may be probed to the same depth. The whole equipment packs into a wooden box (gross weight 38 kg) easily transportable by car. When using the probing tool, the number of blows of the hammer to drive the rods

Fig. 2.1 Some basic items of field equipment.
Back row: Geological hammer and chisel, field notebook, 30-m tape and arrows.
Middle row: Camera, flash gun, compass-clinometer, field glasses (8 × 30).
Front row: Feeler gauge, penknife, scale, hand lens (×10).

Table 2.1 Use of penknife, acid bottle and hand lens

Hardness	*Reaction to 50 per cent HCl*	*Hand lens (×10)*
QUARTZ not scratched by penknife	LIMESTONE, CHALK and MARBLE effervesce vigorously	Identify medium-grained rocks, e.g. SANDSTONE
CALCITE when crystalline, scratched by penknife but not by fingernail (chalk is softer)	MAGNESIAN LIMESTONE and DOLOMITE effervesce mildly	Distinguish between OOLITIC and other limestones
GYPSUM scratched by fingernail	OTHER ROCKS do not effervesce	Examine soils to see grain size and composition, e.g. MICA
Test ROCKS with penknife		Examine rocks for PYRITE
See Chapter 5 for use in identification of rocks		

Fig. 2.2 Boring and prospecting tool. (Courtesy of ELE International Ltd.)

successive depth increments of 150 mm can be used as an approximate indication of the soil's relative density or strength as shown in Table 2.2. The data in Table 2.2 should be considered as a rough guide only, and you should carry out your own calibration of the probe for each particular site as soon as possible, say by comparing the Mackintosh probe results with standard penetration test (SPT) results (see Chapter 8). The tool is particularly useful for applications such as the rapid determination of a sub-alluvial valley profile because of its ability to penetrate soft material rapidly and to give a clear indication of when harder underlying material is reached.

2.2 Abney level

The Abney level is a small hand-held spirit level that is used optically to measure angles from the horizontal. It therefore provides a means of directly measuring the angle of slopes, both natural and artificial. The Abney level is shown in Fig. 2.3 and its use is illustrated in Fig. 2.4. Over short and medium distances a ranging rod with a target fixed at the same height as the observer's eye should be used to sight on; alternatively an assistant of comparable height to the observer can be used to sight on if this is more convenient. Over long distances the offset can be estimated with little loss in accuracy. To use the instrument,

Table 2.2 Penetration resistance with Mackintosh probe

Sands		*Clays*	
Number of blows to drive 150 mm	*Relative density*	*Number of blows to drive 150 mm*	*Strength*
0–10	Very loose	0–5	Very soft
10–25	Loose	5–10	Soft
25–75	Medium dense	10–20	Firm
75–125	Dense	20–40	Stiff
Over 125	Very dense	40–75	Very stiff
		Over 75	Hard

Note: When driving the Mackintosh probe the average energy per blow delivered by hand is 47 J and the values given above have been calculated on this basis.

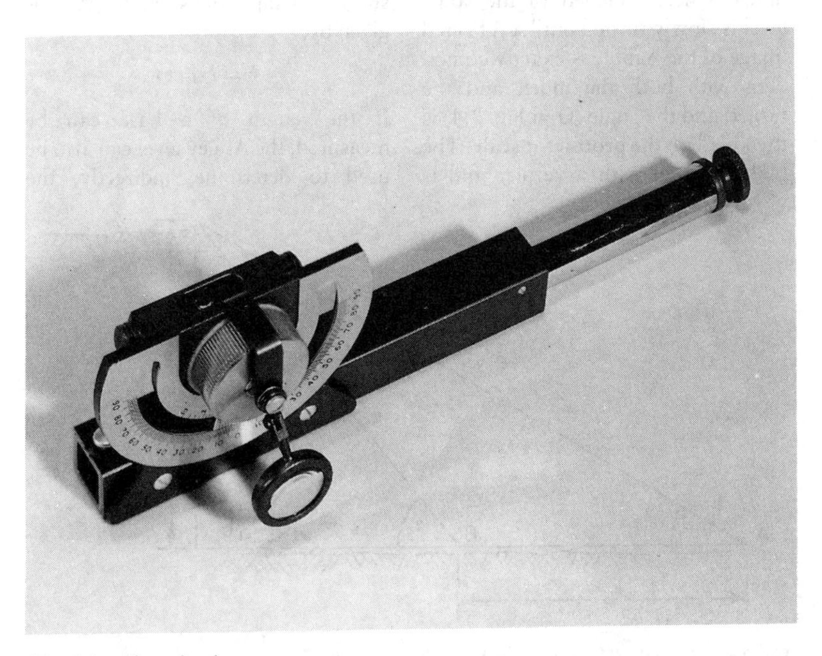

Fig. 2.3 Abney level.

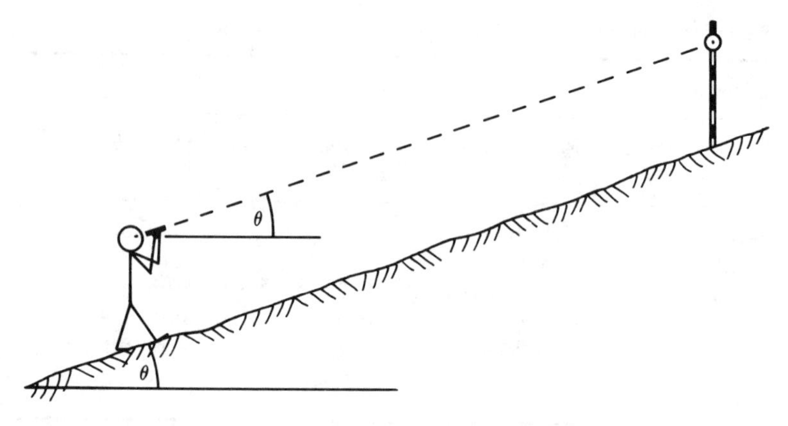

Fig. 2.4 Use of Abney level to measure slope angle Θ.

the index in the sighting tube is first aligned precisely with the target, the milled wheel attached to the spirit level is slowly turned until a reflected image of the bubble is exactly coincident with both the index and the target, and the angle (Θ in Fig. 2.4) is then read on the protractor scale. The scale is fitted with a vernier and is readable to 10 minutes of arc. If the slope length l is measured with a surveyor's tape, the slope height h is given by:

$$h = l \sin \Theta$$

If the length of a base can be measured, the Abney level can also be used to determine, indirectly, the

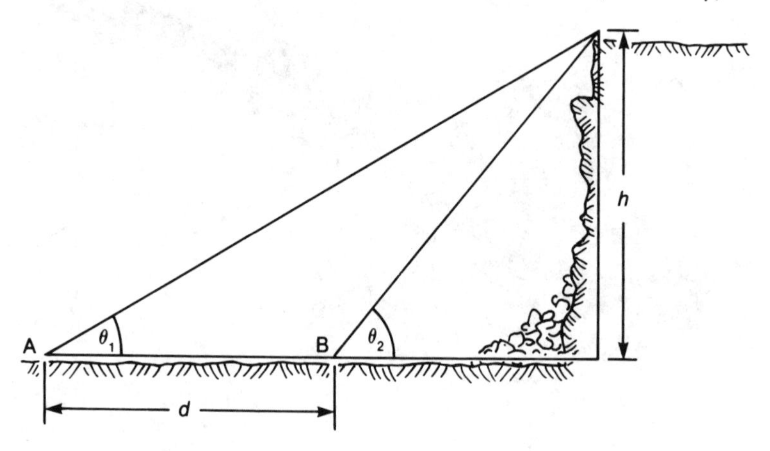

Fig. 2.5 Using the Abney level to determine height h.

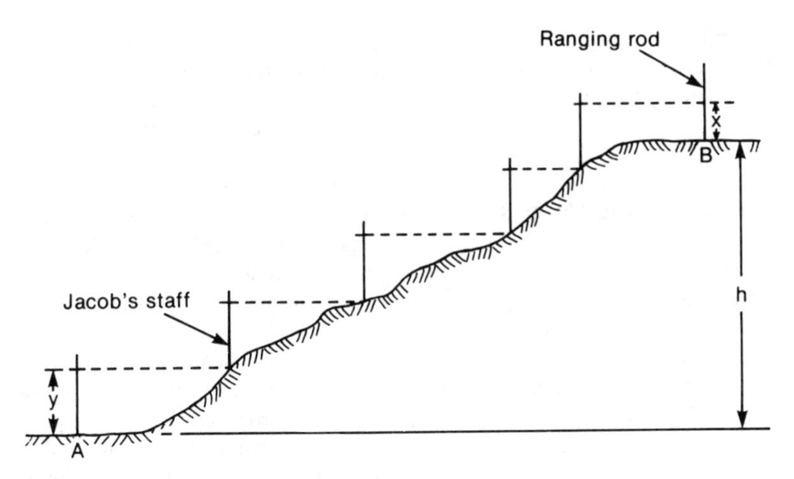

Fig. 2.6 Using Jacob's staff to determine height of a slope.

height of inaccessible vertical features such as cliffs, quarry faces etc. Figure 2.5 shows how the height of an inaccessible point may be determined by observing the angles Θ_1 and Θ_2 with an Abney level at points A and B positioned a measured distance d apart. The height h is given by:

$$h = \frac{d \tan \Theta_1}{1 - \dfrac{\tan \Theta_1}{\tan \Theta_2}}$$

A *Jacob's staff* is an instrument consisting of a ranging rod to which an Abney level is attached so that the sight line of the Abney level, when set on zero, is precisely at right angles to the rod. The point of attachment is such that the sight line of the Abney level is an exactly known distance (called the staff length) from the foot of the rod. A convenient staff length is 1.5 m. The Jacob's staff can be used to determine the height of a regular or irregular slope as shown in Fig. 2.6. The Abney level is set on zero and the staff is positioned at the bottom of the slope on station A. Holding the rod vertical by means of the Abney level bubble, the observer carefully notes the position at which the horizontal sight line intersects the slope (the position can be marked by an assistant if one is available). The staff is then moved to this position and the procedure repeated. The observer records the total number of steps required to reach the top of the slope. If the final step gives rise to a sight line above the top of the slope, the height of intersection of the sight line with a ranging rod held on the final station B is measured (x). For a staff length of y, the height of the slope h in Fig 2.6, where there are five steps, is given by:

$$h = 5y - x$$

A Jacob's staff can also be used as a

convenient means of holding the Abney level steady when using it in the ordinary way to measure angles from the horizontal.

2.3 Prismatic compass

It is assumed that the reader will be familiar with the ordinary geologists' compass-clinometer, the use of which features prominently in the Handbook *Basic Field Mapping* in this series. However, the engineering geologist will also find the prismatic compass extremely useful for particular applications which will be referred to later, so that a description of the instrument will now be given. The prismatic compass allows a precise determination to be made of the magnetic bearing of an object with respect to the position of the observer. The instrument is shown in Fig. 2.7. To use the prismatic compass the lid containing a foresight is opened at right angles to the base and the prism incorporating a backsight is erected as shown in Fig. 2.7. The height of the prism is then adjusted so that the image of the compass card when released is sharp in the eyepiece. The compass card is released onto its pivot and the instrument is brought up to the observer's eye using a thumb through the ring attached to the base to keep it steady in the hand. The foresight and backsight of the com-

Fig. 2.7 Prismatic compass.

pass are brought into line with the object whose bearing is to be determined, and the bearing is then read off the compass card in the prism eyepiece after the compass card has settled to a steady reading. Dry prismatic compasses have a damping pin which allows the observer to damp out large swings of the card and bring it to rest quickly, but liquid ones are self-damping. Prismatic compasses have a card graduated and easily readable to 1°. After use the card should be lifted off the pivot to prevent damage and wear, and the prism and lid folded down. As well as for determining the bearing of an existing object, the prismatic compass can be used to position ranging rods or pegs along any desired bearing. To do this, the observer holds the instrument up to his eye in the previously described manner and slowly turns until the compass card seen through the prism indicates the desired bearing. The compass is now held steady on this bearing. By hand signals, the observer then instructs an assistant holding a ranging rod to move until the rod is coincident with the foresight of the compass. Used in this way together with a surveyor's measuring tape, many simple setting-out tasks for field work in rural areas can be carried out quickly and reasonably accurately. Remember, when using a prismatic compass in conjunction with a map, to make the appropriate correction for the difference between magnetic north and the grid north of the map, this information being provided in the key to the map: if magnetic north is west of grid north the difference must be deducted from, and when magnetic north is east of grid north the difference must be added to, the magnetic bearing in order to give the grid bearing. You can check that you have got this right by measuring with the compass the magnetic bearing of a object whose grid bearing you can determine from the map.

When using the prismatic compass, and geologists' compasses described in Section 2.4, make sure you have no magnetic metal about your person or nearby enough to affect the compass.

2.4 Geologists' compass-clinometers

2.4.1 Suunto compass

The Finnish Suunto geologists' compass (Model G-72) is popular with geology undergraduates and, modified as described below, is suitable for engineering geological field-work. It is a very robust liquid-damped compass fitted with a rotatable cover-window carrying an index mark which registers against the fixed azimuth circle of the compass. The circle is divided in 5° intervals but is readable to 1°. In addition, the instrument has a liquid-damped pendulum clinometer for measuring dip angle, also graduated in 5° intervals and readable with care to 1°. In order to measure strike direction the Suunto compass must be fitted with a round spirit-level as shown in Fig. 2.8. A suitable spirit level can be obtained from a tool

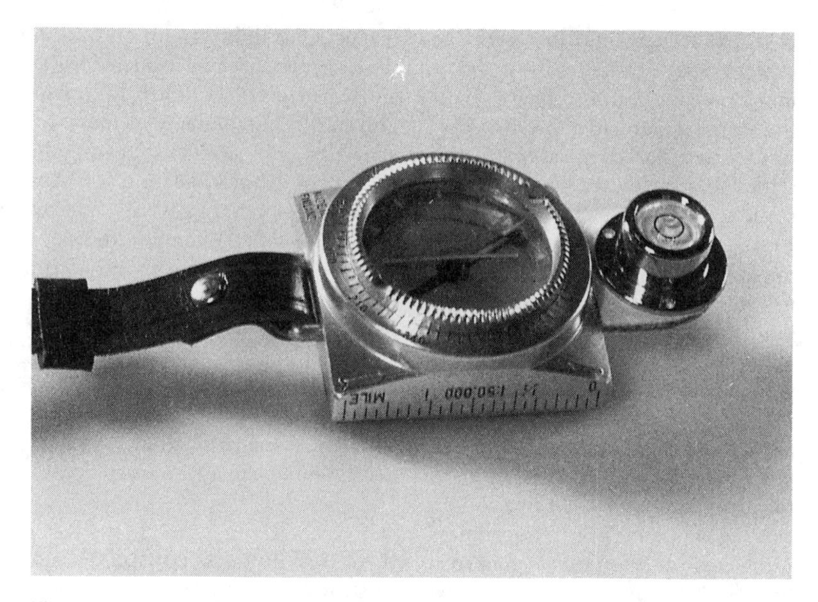

Fig. 2.8 Suunto geologists' compass fitted with spirit-level.

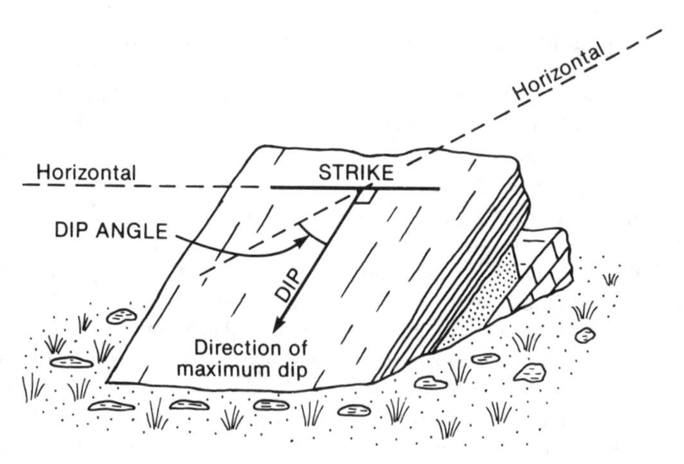

Fig. 2.9 Rock outcrop showing strike, dip, and dip angle.

merchant and fixed to the compass body by means of a small plate made from a non-magnetic material. To measure the strike and dip (Fig. 2.9) of a rock using the Suunto compass proceed as follows:

1 Locate an exposed bedding-plane surface of the rock and position yourself so that the rock dips down to your right.
2 Hold the compass in a horizontal plane using the spirit-level and fit

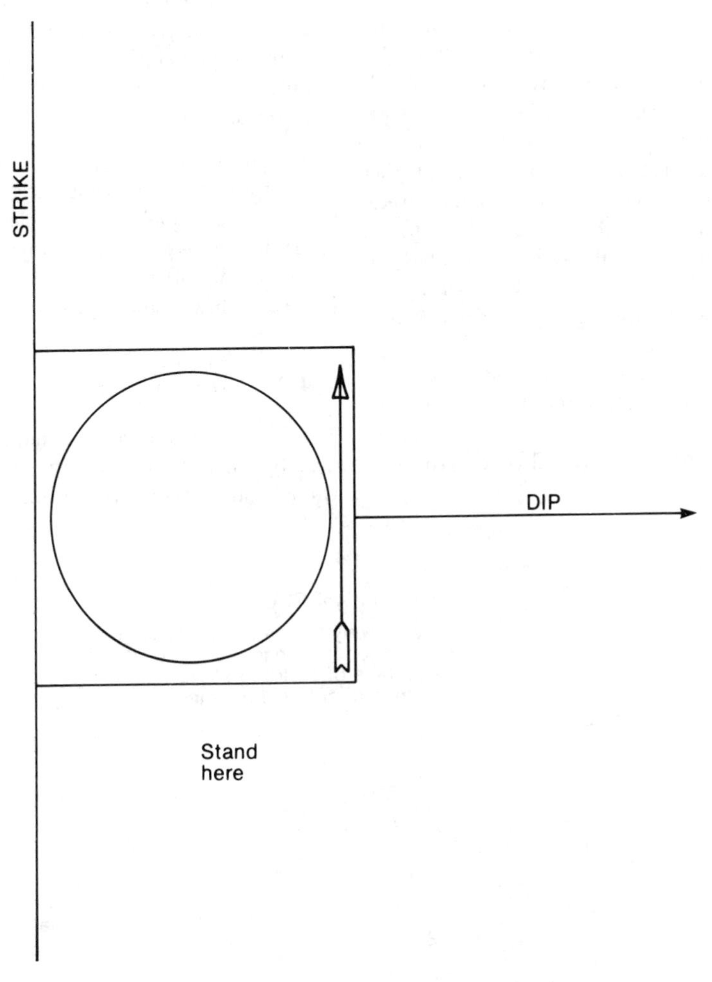

Fig. 2.10 Measuring strike direction.

the left-hand edge of the compass to the rock surface, keeping the compass horizontal.

3 Using the left-hand edge of the compass as a ruler, draw with a pencil a horizontal line on the rock surface. This is a *strike* line (Fig. 2.10).

4 Turn the cover-window until its two luminous marks are directly over the luminous mark on the red end of the compass needle. Read off the bearing on the azimuth circle against the index mark on the cover-window. This is the strike direction. Always record this bearing as a three-figure number to avoid any confusion with dip which will be a one- or two-figure number.

5 Using the compass body as a set square, draw another pencil line at right angles to the strike line going in the down-dip direction. This is a *dip* line.

6 Place the clinometer exactly on the dip line and measure the amount of dip, tapping the instrument once or twice to make sure the pendulum is free.

7 Holding your field notebook so that the top is to the north, record your observations as shown in Fig. 2.11; the diagram should be the same as the pencil lines you made on the rock surface when making the measurements.

Although the example given here has been of a bedding-plane surface, the technique is equally applicable to other surfaces, such as measuring the strike and dip of the surface of a slip plane or a shear plane.

2.4.2 Brunton compass

The Brunton compass is commonly used by engineering geologists in the United States. It consists of a compass

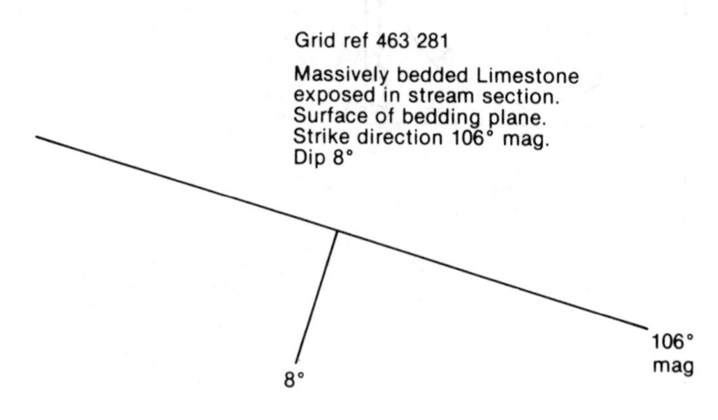

Fig. 2.11 Recording strike and dip in field notebook.

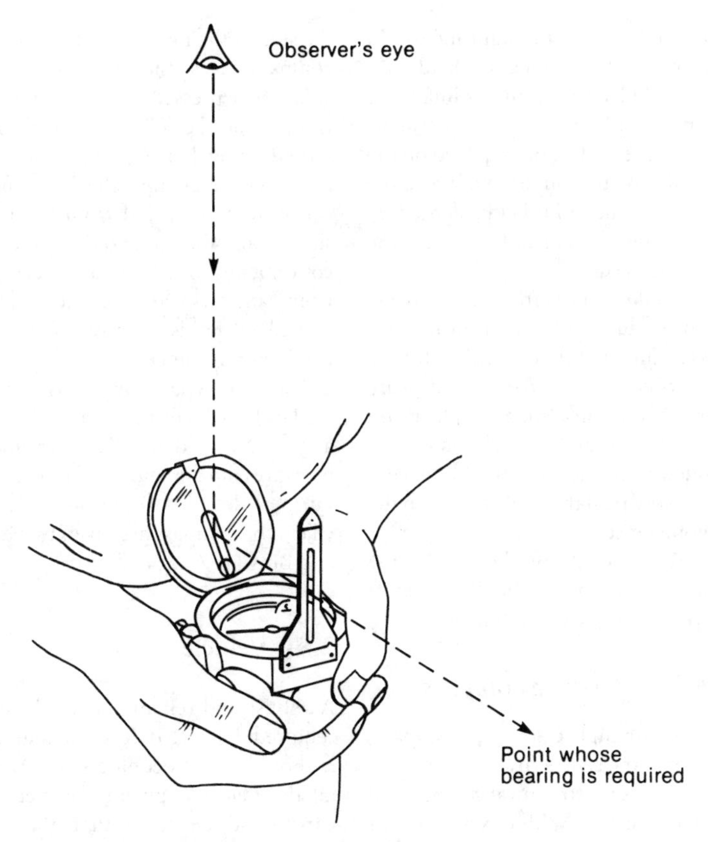

Fig. 2.12 Sighting a distant point with the Brunton compass. (Adapted from Brunton Company illustration.)

having a hinged lid containing a mirror (on which is an axial line) fitted at one end, and a hinged sighting arm fitted at the other end. The compass body itself is fitted with two spirit levels, one circular-bubble for levelling the compass, and one tube-bubble attached to a clinometer.

To take a bearing with the Brunton compass (Fig. 2.12), the lid is opened out to 135° and the sighting arm is turned up so that it is normal to the plane of the compass. Standing with the feet apart, the compass is held at waist height cupped in one hand. The circular bubble is centred and, looking vertically down, the mirror is adjusted so that you can see the point being sighted and the sighting arm in the mirror. The compass is held level and rotated until the mirror images of the point being sighted and the sighting arm are both superimposed on the axial line on the mirror. Now read the bearing of the point being sighted by noting the reading indicated by the compass needle.

To make a measurement of strike, the procedure with the Brunton compass is similar to that described for the Suunto compass above. To measure dip, the Brunton compass is placed on the dip line and the clinometer is rotated until the tube bubble is centred. Now read the angle of dip on the clinometer scale.

Further details of the use of the Brunton compass in field geological work is given by Compton (1962).

2.4.3 Clar-type compass

The Breithaupt Clar-type compass, shown in the centre row of Fig. 2.1, is made for use in structural geology and rock mechanics. With this instrument the dip and dip direction of a rock plane can both be measured in the one operation. The Clar-type compass consists of a compass body containing a circular spirit level and fitted with a hinged lid having a protractor scale engraved on the side of the hinge.

To make a measurement, the hinged lid of the compass is placed against the rock plane to be measured, and the compass body is adjusted until the bubble in the spirit level is centred. The dip of the plane is now given by the reading on the scale on the side of the hinge. Still holding the compass level, the compass needle clamp is depressed thus releasing the compass needle. When the needle has settled, it is locked in position by releasing the clamp. The instrument can now be removed from the rock surface and the dip and dip direction conveniently read, away, perhaps, from an awkward position. (The hinged lid holds its position by the friction of the hinge.)

The Clar-type compass is graduated to 1° in dip direction and 5° in dip angle. Note that the instrument measures dip direction and not strike. Further details of the use of the Clar-type compass are given by Hoek and Bray (1977).

2.5 Vehicle

A robust and reliable field vehicle is required (Fig. 2.13). It is essential that it should have adequate stowage for all the field equipment you need to carry, and desirable that the rear compartment can serve as a small office for dealing with maps, papers, etc. in bad weather. It is useful to have a roof rack for carrying ranging rods, etc. and also helpful if the roof or bonnet is strong enough to stand on so as to provide a vantage point for photography. Four-wheeled drive is essential, and a desirable extra is a winch on the front to get out of very soft or boggy spots where even four-wheel drive may not suffice. The vehicle and all its doors should be securely lockable. A first aid kit must

Fig. 2.13 Vehicle for use on site or in the field.

be carried and everyone should know where it is. The proper beacons and signs should be carried and displayed when working on the public highway. A two-way radio for keeping in contact with base is useful and is a relatively inexpensive item these days.

2.6 Field clothing and safety

Finally, the advice given in *Basic Geological Mapping* on field clothing is sound; the only additions the engineering geologist will need to make are to add a hard hat and boots with protective toe caps when working on engineering sites as discussed later. Appendix 1 and Section 2.1 from the same book should also be studied for the sound advice on safety in the field but further amplification of this aspect of site work will be made in this book at the appropriate places. Wear goggles when using the geological hammer, and keep acid away from skin and eyes when testing rocks for limestone.

- Engineering geologists in Great Britain should always comply with the safety requirements specified in the various British Standard Codes of Practice referred to in this Handbook, particularly those in *CP 2004:1972*.

Fig. 2.14 Map case for use in the field – an example of Government surplus kit.

- Engineering geologists in the United States should always comply with the safety requirements specified by the US Department of Labor for the Construction Industry. In particular they should comply with the Occupational Safety and Health Administration's Standard *OSHA 2207*.

Before leaving the subject of equipment in general, it is worth noting that many useful items can be obtained relatively cheaply from Government surplus stores. Often these include purpose-made pieces such as the map case illustrated in Fig. 2.14.

3
Maps, plans and air photographs

First a word about *scale*. The scale of a map is the ratio of the length of the representation of an object on the map to its actual length on the ground (e.g. 1:63 360 or 1 inch to 1 mile). A *small scale* map is one where this ratio is small and a *large scale* map is one where the ratio is large. Large scale maps such as are used by engineers to show the details of civil engineering sites, usually of scales 1:2500 and larger, are referred to as *plans*. Confusingly, the ruler-like device used to measure distances on maps and plans is also called a 'scale' (see Fig. 2.1). You must have a scale for each scale of map or plan you are using. Most are marked along each edge on both sides giving four graduations on the one scale. For maps a useful scale is one with 1:50 000, 1:25 000 and 1:10 000, but in Great Britain you will need 1:63 360 and 1:10 560 (6 inches to 1 mile) as well. For plans a useful scale is one with 1:2500, 1:500 and 1:100. Since metrication, Imperial scales have been available with metric graduations: these are very convenient when using the appropriate maps but should be marked clearly '*metres*' to prevent mistakes.

In the United States the basic scales you will need are 1:62 500, 1:50 000, 1:24 000 and 1:25 000 for the '15 Minute' and '7.5 Minute' maps.

Advice on the availability of topographical and geological maps in several parts of the world is given in Barnes's *Basic Geological Mapping* in this series. For the United Kingdom you should get hold of a copy of *TRRL Report LR 403 (revised edition)*; it is now ten years since it was last revised but you can soon bring it up to date by making your own annotations and additions.

3.1 Topographical maps

For work in Great Britain, the 1:25 000 scale map is very useful for the preliminary assessment and inspection of a site. It is the largest scale Ordnance Survey map series printed in colour, although there is also an outline edition. The colour on the map is of considerable help in making a first appraisal of the site, especially the use of blue to mark water courses, areas of water, marshes, and their associated names. Springs are also shown. A regular pattern of straight artificial drainage channels is a good indication of a high water table. Field boundaries are shown and there is good indication of the type of vegetation in woodlands and uncultivated ground. The contour interval is 25 feet, which can help in the picking out of steep ground which may be liable to instability. Public footpaths, bridle paths, and other rights-of-way which may be useful for getting access to the site, are clearly marked on the more recent revisions of the England and Wales sheets of the 1:25 000 map.

The 1:10 560 scale maps cover the whole of Great Britain. The Provisional Series has a 50-ft contour interval (sometimes 100 ft), and the Regular Edition has 25-ft contours. As part of the metrication programme the 1:10 560 scale maps are being replaced by 1:10 000 maps with a contour interval of 5 m (10 m in the more mountainous areas). This is the largest scale of Ordnance Survey map to show contours and the smallest on which features are generally shown to correct scale.

For general planning and location purposes small scale maps may be useful. The once familiar 'One-Inch' (1:63 360) Ordnance Survey maps have now been replaced by the 1:50 000 'Landranger' series which covers the whole of Great Britain. The Route Planner Map (north and south sheets, scale 1:625 000) shows motorways, primary routes, and many other roads, and shows which have dual carriageways; a new edition is published annually.

3.2 Geological maps

The British Geological Survey's 1:63 360 series of geological maps (with memoirs) gives a good idea of the types of material and the structures occurring in the locality. 'Solid' or 'Drift' editions are generally available. The 1-inch sheets are being replaced by 1:50 000 scale sheets using the same sheet outlines. For more detailed geological information on the site, the 1:10 560 scale maps – now being replaced by 1:10 000 sheets – should be examined; they often have descriptive notes printed on them, and the positions of many boreholes are marked with brief borehole data. The Survey has also published several geological maps relating to new towns and classic geological areas at a scale of 1:25 000, some with explanatory booklets.

For planning purposes, maps of a smaller scale may be useful, and there are two versions of the 1:625 000 (about 1 inch to 10 miles) map showing either the Solid or the Quaternary geology of the whole country in two sheets (north and south).

Up-to-date information on the current cover of Ordnance Survey and Geological Survey maps for Great Britain can be had from the Ordnance Survey *Trade Catalogue*, published annually, and available from the Ordnance Survey, Romsey Road, Maybush, Southampton, SO9 4DH.

3.2.1 US topographical and geological maps

In the United States the most convenient maps to work with will be the 'Standard Series of Topographical Maps' and the 'Geological Survey Maps'. Both are available in the '15 Minute Series' to a scale of 1 : 62 500 (or 1 : 50 000) having contours and elevations shown in feet (or metres), and the '7.5 Minute Series' to a scale of 1 : 24 000 (or 1 : 25 000) having contours and elevations shown in feet (or metres).

Information on these Series of maps can be obtained from the US Department of the Interior, Geological Survey National Center, Reston, Virginia.

3.3 Plans

Civil engineering plans of the site may be available at a number of scales. For motorway sites, for example, plans are generally at 1 : 2500 or 1 : 1000, but for areas of particular detail there may be plans at 1 : 500 or larger as well. Where there is a choice of scale, choose the plans at a scale most suited to your purposes, for example a smaller scale for a general soil survey, or a larger scale for the mapping of excavations.

Plans will normally constitute the base maps on which you will record your observations on soils and rocks and the positions of exposures, trial pits, trenches, boreholes, etc. (Fig. 3.1). They should also be used to record any physiographical information not shown on existing topographical maps such as abrupt changes in slope, instability, springs and seepages, swallow holes, river terraces, made ground, etc. The techniques of mapping are outside the scope of this *Handbook* but a great deal of the advice given in *Basic Geological Mapping* will be directly applicable, and there is a wealth of information given by the Geological Society Engineering Group Working Party (1972).

Topographical and geological field survey methods are outside the scope of this *Handbook*, but much useful advice on these matters is given by Lahee (1961).

When working in the field, always keep your maps or plans in a waterproof map case (see Fig. 2.14 for suitable example).

3.4 Air photographs

For working in the field with air photographs some basic equipment is needed. First is a pocket stereoscope; buy a good quality model fitted with *optical-quality glass lenses* and on which the distance between the lenses can be varied to suit the eye-base separation of the user. The second is a viewing board; this can be homemade from waterproof plywood (dimensions 255 × 460 mm are suitable for standard 230-mm square air photographs) and should have battens fitted underneath to give clearance for bulldog clips which hold the photographs in position on the board. The third is a viewing aid. The viewing aid (Fig. 3.2), which can be readily made from sheet metal, is not essential, but makes working in the field very much easier by accommodating the bent-up inner edge of the overlapping upper photograph of a stereo-pair. Use of the viewing aid has a number of advantages: it makes the whole area available for examination, it prevents one eye from straying into the field of view of the other, it prevents the wind from catching the

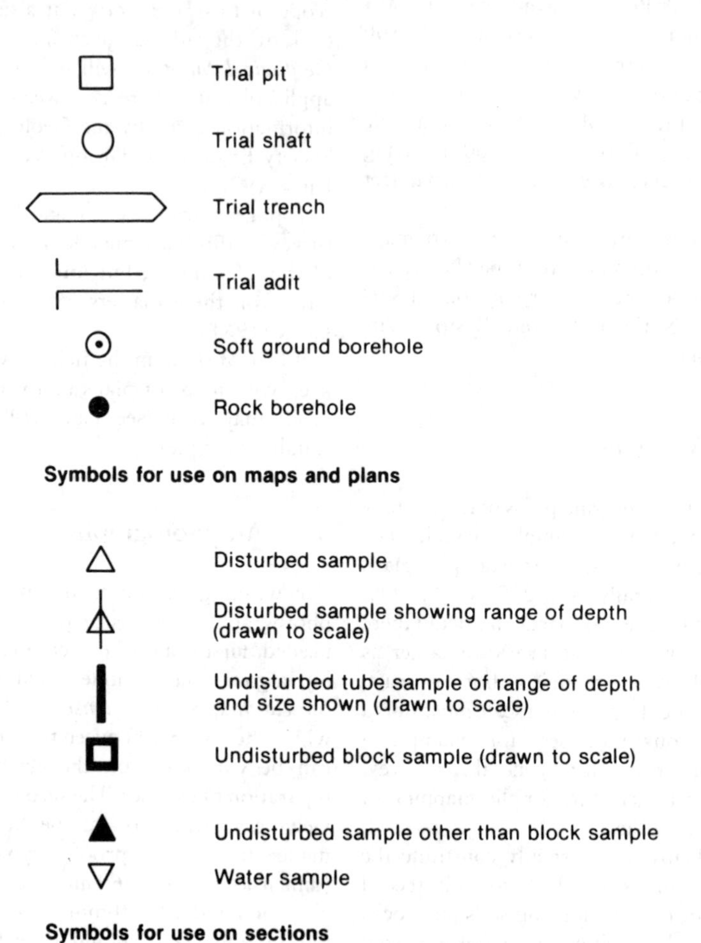

Fig. 3.1 Symbols for use on maps, plans, sections and field notebooks. Write the reference number alongside the symbol.

overlap and it leaves the hands free from turning up the overlap. The viewing aid is also held down on the viewing board with bulldog clips. Finally *Chinagraph* pencils, in different colours, are needed to annotate the photographs and some cotton wool for rubbing out the annotations if required. Only alternate photographs need be marked, because a line drawn on one photograph of a stereopair will be seen clearly in the stereoscopic image. When in the field, the viewing board and stereoscope can be carried on cords round the neck (Fig. 3.3).

For new civil engineering works in Britain two sets of air photographs are often taken for the topographical survey, one at a scale of about 1 : 10 000 for the preliminary survey and another at 1 : 3000 for the detailed site plans. The 1 : 10 000 scale photographs will be useful for making a reconnaissance of the site by car but the 1 : 3000 set will be more useful for working with in the field on foot. If an index plot showing the outline of

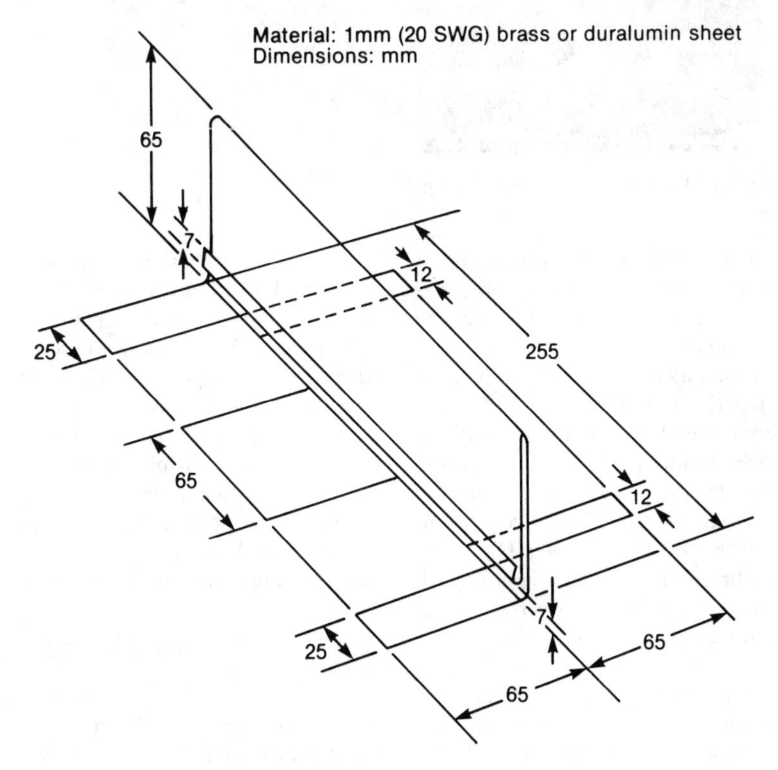

Fig. 3.2 Viewing aid for air photographs.

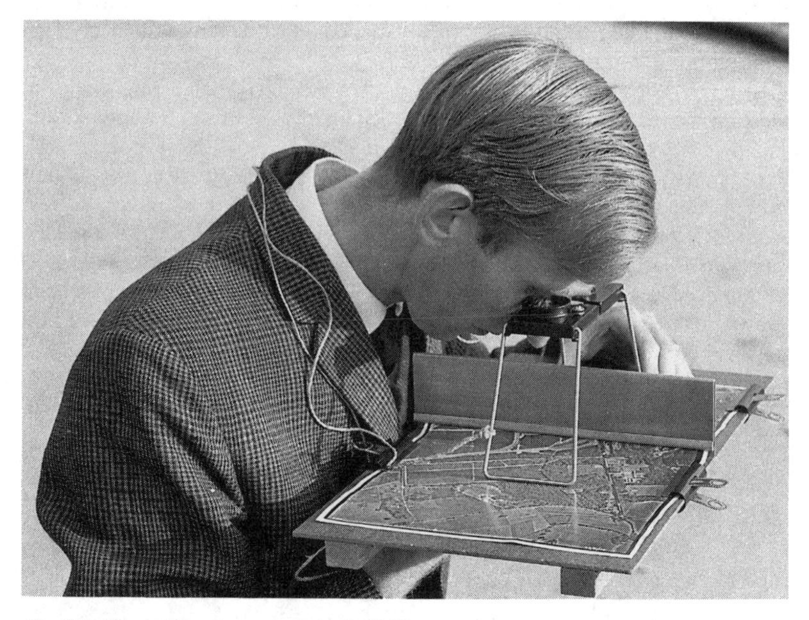

Fig. 3.3 Examining a stereopair in the field.

alternate photographs on a map is not supplied with the photographs, it is essential to make one so that the stereopairs covering a particular area can be quickly found. A template in the form of a square frame of the right size is useful for drawing the outlines, each of which is identified by marking it with the serial number of the photograph. A quicker procedure is to plot the flight strips, identifying only the first and last photographs of each flight. Some basic properties of air photographs are explained in Section 3.4.3.

It is assumed that the reader will know how to work with stereoscopic pairs of air photographs and will also know some basic photo-geological interpretation. If not, read Sections 3.4.3 and 3.4.4 of this *Handbook* and Sections 3.7 and 4.9 of *Basic Geological Mapping*. What follows is some advice that the engineering geologist may find useful when using air photographs in the field. There are two principal uses for air photographs in the field: the first is their use as a supplement to topographical maps and the second is their use as a source of information on ground conditions.

3.4.1 Air photographs as a supplement to maps

Air photographs provide more detailed information on the area than can be given by a map. A map can represent only a small amount of the features present in the area it covers, and that only in symbolic form. Examination of the photographs will give information on the character of the features represented symbolically on the map, as well as additional information not represented on the map, e.g. character of boundaries (hedges, ditches, fences, etc.), character of buildings, width of streams, roads and verges, position and height of trees, density of woods, location of poles and pylons supporting wires, and agricultural use of fields. Features which have appeared since the map was last revised will also be shown on more recent photographs.

In the field, in areas where the map shows little detail, the photographs may be a valuable aid in locating one's position on the ground by reference to small unmapped features such as trees, bushes, gullies, wadis or rocky outcrops. The photographs can save much time by indicating routes of access, such as the position of gates and gaps in hedges. They may also be used in the field as a base map on which to mark the position of observed features or sampling points. In undeveloped countries where topographical maps may be rudimentary, or may not even exist at all, air photographs will be invaluable. In addition to providing a base map of the area they may be the only means of finding your way about in the field.

Old air photographs, which can be obtained for some areas dating back to as long ago as the 1920s, form a record of past conditions. They will be particularly useful when examining a site which is now covered with more recent urban or industrial development.

3.4.2 Interpretation of ground conditions from air photographs

In addition to their use for giving information on surface features, air photographs may also be used to draw conclusions on ground conditions of engineering significance, such as soil type, drainage, marshy areas, unstable ground, mining subsidence, swallow-holes, spring lines and rock outcrops.

The type of features likely to occur in the particular terrain should be determined from the general geology of the area, and the air photographs should be examined with these features in mind. Relief is an extremely valuable factor in the interpretation of air photographs. The steepness of a slope tends to be characteristic of the material on which it has formed, so that a change in material is often marked by a change in slope which can be detected in the exaggerated relief obtained with the stereoscope. On black-and-white photographs all detail and form, including differences in colour, are represented by patterns and textures in tones of grey, and tone may offer some guide to materials when they are exposed at the surface. Dark tones are often shown by moist soils, or where the drainage is impeded as in waterlogged and marshy ground and in unstable areas. Light tones are often shown by well-drained soils. Standing water usually appears dark where direct reflections are absent, and so do water-seeps and spring lines. Coarse-grained and rocky materials usually give rise to a rough surface texture, and fine-grained material to a smooth one, except where the surface has been disturbed by slope instability when a roughened or hummocky texture may be seen.

Air photographs have been particularly useful in locating areas of landslip and the inexperienced engineering geologist is strongly urged to examine some stereo-pairs of the classic examples. When working in the field anywhere on sloping ground, but particularly along the scarp slopes of escarpments, the photographs and the ground should be carefully examined for signs of instability. (See also Section 7.2.1 for ground-based stereo-photography).

Engineering geologists working abroad are warned that in some countries air photographs are regarded as security material and attempts to obtain them should first be cleared with the appropriate Government Department.

3.4.3 Some properties of air photographs

For engineering geologists who have not worked with air photographs before, some simple properties of air photographs will now be described.

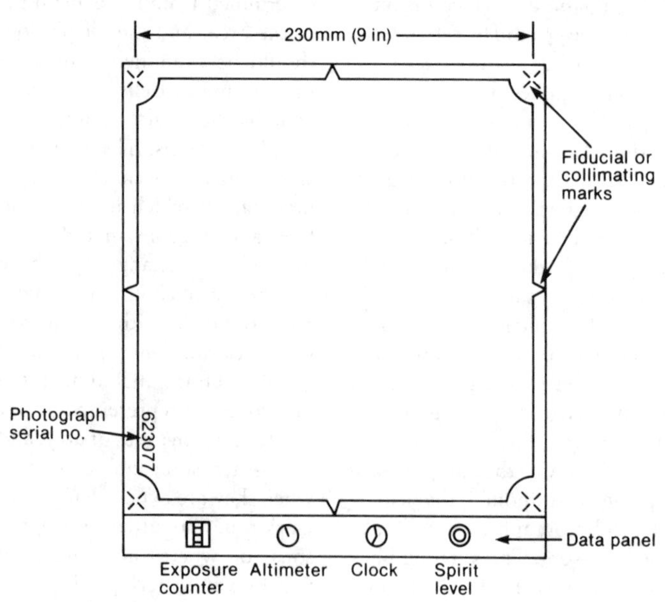

Fig. 3.4 Format of a typical air photograph.

The format of a vertical air photograph is shown in Fig. 3.4. If the data panel is preserved it will show photographs of a spirit level, a clock, an altimeter and an exposure counter. The focal length of the camera lens, the date of the photography, the sortie, run and film reference numbers, and the serial number of the photograph may also be shown. A 230-mm square photograph format is standard, and most modern air survey cameras have a 152-mm focal length lens. The scale of a photograph, s, can be obtained from the formula $s = f/(H-h)$ where f is the focal length of the camera lens. H is the height of the camera above sea level, and h is the height of the ground above sea level (see Fig. 3.5). The scale can also be obtained by comparing the distance between two features on a photograph with the distance between the same pair of features on a map of a known scale. The points should be of similar height above sea level. Photographs are often supplied with the data panel trimmed off but with the scale included in a title written along the border. Scales are nominal, because the scales to which features are rendered in the photograph vary with the relief, i.e. with the distance from the camera. Some air photographs taken with multiple camera installations are not truly vertical, so that the scale is not constant across the photograph.

In an air survey a strip of ground is photographed by an aeroplane flying at constant height on a straight course. The speed of the aeroplane and the interval between successive photographic exposures are so arranged that the field of each photograph overlaps that of the previous photograph by about 60 per cent, and this ensures that each point on the

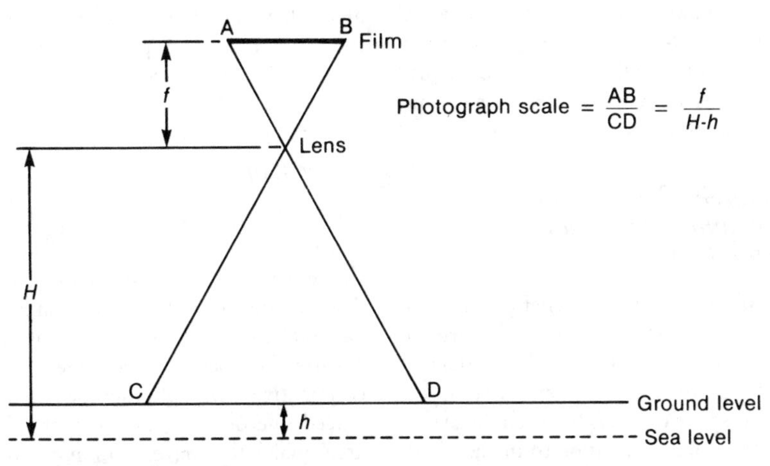

Fig. 3.5 Scale of an air photograph.

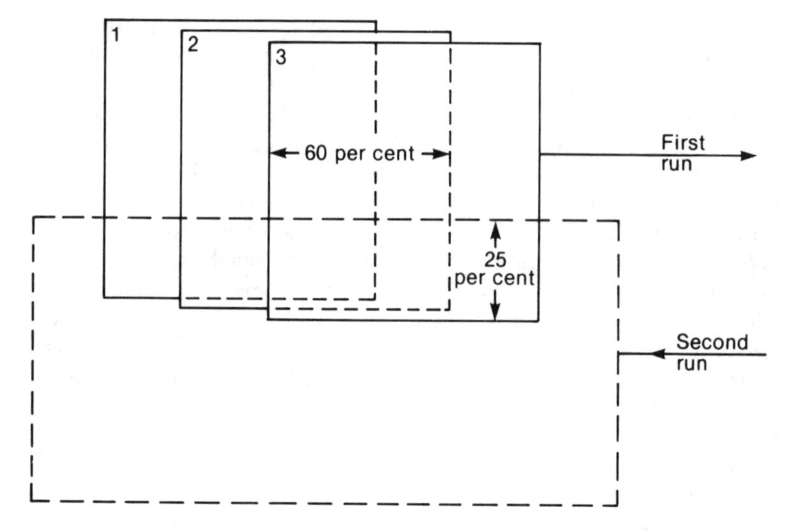

Fig. 3.6 Overlap of air photographs.

ground appears on at least two photographs (Fig. 3.6). To cover a broad area of land a number of parallel strips are flown with a lateral overlap of about 25 per cent. The area covered by a 230-mm square air photograph in relation to the 1-km National Grid lines is shown in Fig. 3.7 for various scales of photography.

3.4.4 Stereoscopic examination of air photographs

Any impression of relief experienced when looking down at the ground from an aeroplane is due to factors other than stereoscopic vision, because the distance between the eyes is very small in relation to the height of the aeroplane. However, the very much greater distance between the stations from which successive air photographs are taken allows the ground to be seen in greatly exaggerated relief when the area common to two consecutive air photographs is examined with a stereoscope.

3.4.5 Using the pocket stereoscope

The pocket stereoscope (see Fig. 3.3) consists of two lenses mounted in a frame which has folding legs. The distance between the lenses can be varied to suit the eye-base of the user. To use the pocket stereoscope, two consecutive air photographs are placed one on the other so that the areas which they have in common are in register. An item of detail is selected

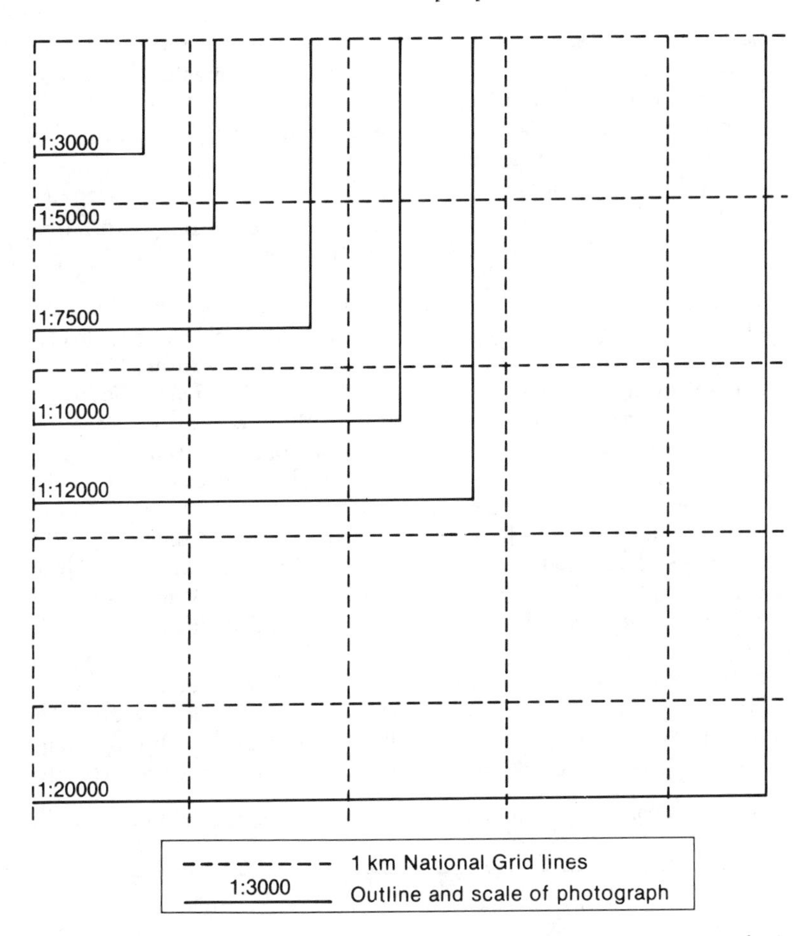

Fig. 3.7 Area covered by 230-mm (9-in.) square air photographs at various scales in relation to the 1-km national grid lines.

in the area of overlap and the photographs are drawn apart until the separation of this item on the two photographs is the same as the distance between the observer's eyes, about 60 mm. A strip of about 60-mm width is now available for stereoscopic examination. The stereoscope is placed with each lens over the same item of detail on the two photographs. A single stereoscopic image should now be seen through the stereoscope.

(At this stage you may find that you need to make small adjustments in the separation of the photographs and in the position of the stereoscope.) It will help if the photographs can be positioned so that any shadows in them appear to fall towards the observer, and if the direction of the light during examination is arranged to correspond with the direction of light in the photographs. It will also help to look first at a stereopair of a bold landscape in which features expected to show relief can be anticipated. Features showing high relief may be difficult or impossible to view stereoscopically, e.g. electricity supply pylons.

A strip of only about 60-mm width can be viewed stereoscopically at one time with a pocket stereoscope. If the overlapping of the photographs is interchanged, the lower being placed on top, another strip of the same width can be viewed stereoscopically, leaving in a standard 230-mm square photograph a strip about 22-mm wide still hidden. To examine this strip it is necessary to bend up the inner edge of the upper photograph. This can be done by hand, or a viewing aid made of sheet metal may be used (see Figs 3.2 and 3.3).

3.5 Sources of maps and air photographs

In Great Britain topographical maps are readily obtainable from Ordnance Survey agents throughout the country, or direct from the Ordnance Survey, Romsey Road, Maybush, Southampton, SO9 4DH. Geological maps can be obtained direct from the British Geological Survey, Nicker Hill, Keyworth, Nottingham, NG12 5GG, and its regional offices; some Ordnance Survey agents also carry stocks of local geological maps. Air photographs are available from a number of sources but, to begin with, enquiries should be directed to the Air Photo Cover Group, Ordnance Survey, at the address given above, or the

Table 3.1 Objectives when examining maps, air photographs and plans of a site

Lithology:	Main rocks and soils present on site, i.e. Solid and Drift geology
Structure:	Attitude of rocks
	Folding
	Faults
Surface:	Physiography (valleys, escarpments, terraces, etc.)
Ground-water:	Aquifers
	Springs, seepage
	Water table
Hazards:	Landslips, slope instability
	Subsidence, old mineworkings, swallow holes
	Contaminated land

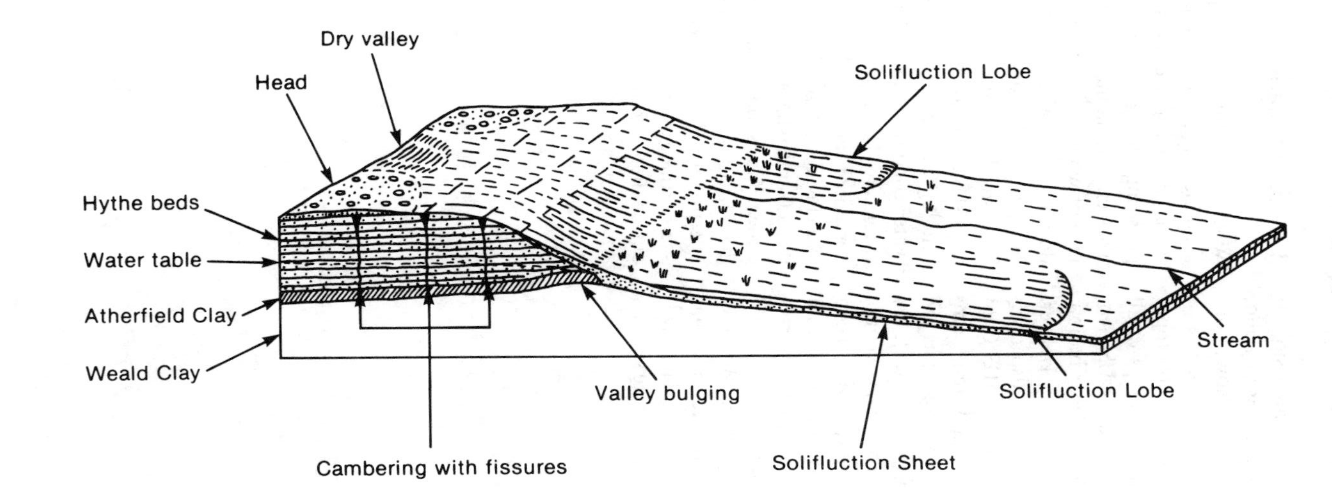

Fig. 3.8 Block diagram of a site.

Air Photo Unit, Royal Commission on the Historic Monuments of England, Fortress House, 23 Savile Row, London, W1X 2JQ.

In the United States both topographical and geological maps of all areas are available from the US Geological Survey, Map Distribution, Federal Center, Box 25286, Denver, Colorado 80225. Maps are also available over-the-counter from US Geological Survey offices and some commercial dealers. Air photographs are available from very many sources, but for information on national cover the US Geological Survey, EROS Data Center, Sioux Falls, South Dakota 57198, and the National Cartographic Information Center, 507 National Center, Reston, Virginia 22092, are the two agencies that should be contacted first. They will direct users to the appropriate sources of air photographs.

3.6 Examination

When examining topographical and geological maps, air photographs and plans of an engineering site, the engineering geologist should proceed in a systematic manner and have a clear idea of what he is looking for. If you do not do this there is the danger of overlooking important items. It is suggested that to begin with the objectives should be to clearly identify the lithology, structure, surface, groundwater conditions and hazards for the whole site. These are itemized in more detail in Table 3.1 which can be used as a checklist. A good discipline to see if you understand the site in three-dimensions is to try and draw a block diagram of it (Fig. 3.8). As a check on interpretation, an inspection of the site on the ground should be carried out at the same time as the examination of maps and air photographs.

4
Description of soils

Engineering soils can be considered to fall broadly into three groups: residual soils, transported soils, and what the geologist would call non-indurated rocks.

- *Residual soils* are those that have formed in place by the direct weathering of rocks. In cold and cool-temperate climatic zones they are usually fairly thin because weathering is slow. In hot and warm-temperate regions residual soils may be thick because weathering is rapid.
- *Transported soils* have been carried to their present location by the action of some natural transporting agent such as water, ice or wind. Examples are the alluvial and estuarine deposits in river valleys and estuaries, and the tills and morainic deposits produced by glacial action.
- *Non-indurated rocks* are often referred to as soils by the engineer. Examples are the Oxford Clay, the Keuper Marl, and some of the Cretaceous and Tertiary sand formations.

If a sample of soil is crumbled in the hand and then examined, it will be seen to consist of particles showing a range of size. The coarse particles (greater than 2 mm) are mainly rock fragments. The medium-sized particles (say 0.1–2.0 mm) are usually mineral particles, often predominantly quartz. The finer material is a mixture of silt and clay that is too fine to identify visually in the field, but which can be assessed using the properties of plasticity and dilatancy described below.

A full description of soils comprises two parts: a description of the soil mass which can only be made from exposures such as trial pits or in a more limited way from undisturbed samples, and a description of the soil material which is made by examination of a disturbed sample of the soil. The description of the soil material leads to the *soil name* and the description of both the soil material and the soil mass leads to the *full soil description*. Both these can be done in the field using simple tests and observations which will now be described.

4.1 Description of soil mass

To make a field description of the soil mass it is necessary to examine an exposure, or failing this, an undisturbed sample. The first thing to do is to record the presence of any bedding using the scale of bedding spacing giving in Table 4.1. It layers of alternating soil types are too thin to be recorded individually the soil may be described as interbedded or interlaminated. Any special bedding characteristics such as cross-bedding or graded bedding should also be recorded. *Discontinuities* such as bedding plane partings, joints, fissures, faults and shear planes should then be recorded using the scale of spacing for these features given in Table 4.1. Particular attention should be given to locating any major slip planes that may be present, especially if the job is an instability investigation. Major slip planes may be overlooked, being taken for bedding plane partings, unless you know what to look for; the inexperienced engineering geologist should read case histories describing the occurrence of slip planes and try, if possible, to examine them in trial pits under the instruction of an experienced colleague. The presence of a polished slickensided parting is one of the key signs to look for both in borehold core and in trial pit exposures. Any other structural features of the soil mass should be recorded. Evidence of weathering should be noted; in silts and clays this is usually shown by a columnar or crumb structure, in gravel and coarser particles by a weakened state or by concentric layering. If appropriate, the terms in

Table 4.1 Interval scales for describing spacing of bedding (left) and discontinuities (right)

Scale of bedding spacing		*Scale of spacing of discontinuities*	
Term	*Mean spacing (mm)*	*Term*	*Mean spacing (mm)*
Very thickly bedded	over 2000	Very widely spaced	over 2000
Thickly bedded	2000 to 600	Widely spaced	2000 to 600
Medium bedded	600 to 200	Medium spaced	600 to 200
Thinly bedded	200 to 60	Closely spaced	200 to 60
Very thinly bedded	60 to 20	Very closely spaced	60 to 20
Thickly laminated	20 to 6	Extremely closely spaced	under 20
Thinly laminated	under 6		

(After BS 5930:1981)

Table 4.2 Field estimation of compactness and strength of soils

		Term	*Field test*
GRANULAR SOILS	BOULDERS AND COBBLES	Loose	By inspection of voids and particle packing.
		Dense	
	GRAVELS AND SANDS	Loose	Can be excavated with a spade; 50 mm wooden peg can be easily driven.
		Dense	Requires pick for excavation; 50 mm wooden peg hard to drive.
		Slightly cemented	Visual examination; pick removes soil in lumps which can be abraded.
COHESIVE SOILS	SILTS	Soft or loose	Easily moulded or crushed in the fingers.
		Firm or dense	Can be moulded or crushed by strong pressure in the fingers.
	CLAYS	Very soft	Exudes between fingers when squeezed in hand.
		Soft	Moulded by light finger pressure.
		Firm	Can be moulded by strong finger pressure.
		Stiff	Cannot be moulded by fingers. Can be indented by thumb.
		Very stiff	Can be indented by thumb nail.
	PEAT	Firm	Fibres already compressed together.
		Spongy	Very compressible and open structure.
		Plastic	Can be moulded in hand, and smears fingers.

(Adapted from BS 5930:1981)

Table 5.6 can be used. Finally the compactness or strength of the soil mass should be estimated. This is done using the simple field tests described in Table 4.2, which also gives the terms to be used. The tests should always be carried out at the field moisture content. For fine-grained cohesive soils the field tests given in Section 6.2 can be used to measure the strength of the soil.

4.2 Description of soil material

To make a field description of the soil material, an examination of a representative sample is carried out by

Table 4.3 Field identification of basic soil types

		Basic soil types	*Particle size (mm)*		*Visual identification*
GRANULAR SOILS	Very coarse soils	BOULDERS			Only seen complete in pits or exposures.
				200	
		COBBLES			Often difficult to recover from boreholes.
				60	
	Coarse soils (over 65 per cent sand and gravel sizes)	GRAVEL	coarse		Easily visible to naked eye; particle shape can be described; grading can be described.
				20	Well graded: wide range of grain sizes, well distributed. Poorly graded: not well graded.
			medium		(May be uniform: size of most particles lies between narrow limits; or gap graded: an intermediate size of particle is missing.)
				6	
			fine		
				2	
		SAND	coarse		Visible to naked eye; very little or no cohesion when dry; grading can be described.
				0.6	
			medium		Designated well or poorly graded as for gravel.
				0.2	
			fine		
				0.06	

Table 4.3—continued

		Basic soil types	*Particle size (mm)*	*Visual identification*
COHESIVE SOILS	Fine soils (over 35 per cent silt and clay sizes)	SILT	coarse — 0.02 medium — 0.006 fine — 0.002	Only coarse silt barely visible to naked eye; exhibits little plasticity and marked dilatancy; slightly granular or silky to the touch. Disintegrates in water; lumps dry quickly; possess cohesion but can be powdered easily between fingers.
		CLAY		Dry lumps can be broken but not powdered between the fingers; they also disintegrate under water but more slowly than silt; smooth to the touch; exhibits plasticity but no dilatancy; sticks to the fingers and dries slowly; shrinks appreciably on drying usually showing cracks.
	Organic soils	ORGANIC CLAY, SILT or SAND	Varies	Contains substantial amounts of organic vegetable matter.
		PEAT	Varies	Predominantly plant remains usually dark brown or black in colour, often with distinctive smell; low bulk density.

(Adapted from BS 5930:1981)

crumbling it or moulding it in the hands. By a visual examination it is first decided whether the soil is granular (tends to break down to individual particles) or cohesive (tends to stay together as a lump). If it is granular, then the sample is crumbled and the range of particle sizes present is estimated. If it is cohesive, the sample is moulded, adding water if necessary, and the plasticity and dilatancy are assessed: plasticity is exhibited if the moist lump can be shaped like Plasticine, and dilatancy is shown if a pat of wet soil is squeezed and surface moisture disappears inside. If it is predominantly composed of plant remains it is classified as fibrous or amorphous peat depending on the presence or absence of recognizable fibres. Table 4.3 gives the criteria to be used for the field identification of soils. Note that the engineering expression *well graded* means a wide range of grain sizes is present and is the opposite of the geological expression *well sorted* which means a restricted range of grain sizes is present. And similarly *poorly graded* is the opposite of *poorly sorted*. Much confusion can be caused unless this difference is realized.

Many soils will be mixtures of the basic types shown in Table 4.3 and it is also necessary to describe these. However, it is recommended that the field description should not try and recognize too many composite soil types and that, in practice, it will be sufficient to use only those given in Table 4.4. Ideally, the soil name should be written out in full with the noun in capitals as shown in Table 4.4, but if a lot of soil descriptions are being made abbreviations can be used. For example, bo, co, gr, sa, si and cl can be used for boulders, cobbles, gravel, sand, silt and clay; using these, very clayey GRAVEL would be abbreviated v cl GR. If abbreviations are used the meanings of them should be clearly listed in the front of the field notebook and this should be repeated each time a new field notebook is used. The experienced engineering geologist may wish to use the symbols of British Soil Classification System (BSCS) shown in Table 4.4. Note that the letters of the BSCS symbols are used in the reverse order to the words of the written description; for example, the symbol for sandy CLAY is CS. The BSCS symbols must be written in brackets when a field, as distinct from a laboratory, identification has been made and they have been put in brackets in Table 4.4 to remind the user of this. If the BSCS symbols are being used this fact should be stated in the front of the field notebook.

Here are some useful tips. The boundary between boulders and cobbles is about the length of a brick, that between cobbles and gravel is about the size of a hen's egg, that between gravel and sand is about the thickness of a matchstick, and that between sand and silt is barely visible to the naked eye. Silts dry on the hands quicker than clays.

Coarse soils may be a useful source of concrete aggregate so remember to describe the particle shape as angular, subangular, subrounded, rounded, flat or elongated. Also note the rock type of the larger particles. Finally, describe the colour of the soils using the colour names red, pink, yellow, brown, olive, green, blue, white, grey and black or, more fully, using the *Munsell Soil Color Charts* supplied by the Munsell Color Co. (1954). A record of the colour of the soils will be particularly useful later on when identifying them in colour photographs of trial pits, exposures, etc. Remember that some soil samples will later be classified fully after plasticity or grading tests have been done in the laboratory. Always try and get these results and compare them with your

Table 4.4 Names for composite soil types

			Soil types and names	*BSCS symbols*
GRANULAR SOILS			GRAVEL and SAND may be designated Sandy GRAVEL and Gravelly SAND	 (GS) (SG)
	COARSE SOILS less than 35 per cent of the material is finer than 0.06 mm	GRAVELS More than 50 per cent of coarse material is of gravel size (coarser than 2 mm)	Well graded GRAVEL Poorly graded GRAVEL	(GW) (GP)
			Silty GRAVEL Clayey GRAVEL	(G-M) (G-C)
			Very silty GRAVEL Very clayey GRAVEL	(GM) (GC)
		SANDS More than 50 per cent of coarse material is of sand size (finer than 2 mm)	Well graded SAND Poorly graded SAND	(SW) (SP)
			Silty SAND Clayey SAND	(S-M) (S-C)
			Very silty SAND Very clayey SAND	(SM) (SC)
COHESIVE SOILS	FINE SOILS more than 35 per cent of the material is finer than 0.06 mm	Gravelly or sandy SILTS and CLAYS 35 per cent to 65 per cent fines	Gravelly SILT Gravelly CLAY	(MG) (CG)
			Sandy SILT Sandy CLAY	(MS) (CS)
		SILTS and CLAYS 65 per cent to 100 per cent fines	SILT CLAY Silty CLAY	(M) (C) –
	ORGANIC SOILS		Example: Organic SILT	(MO)
	PEAT		Fibrous or amorphous PEAT	(Pt)

(Adapted from BS 5930:1981)

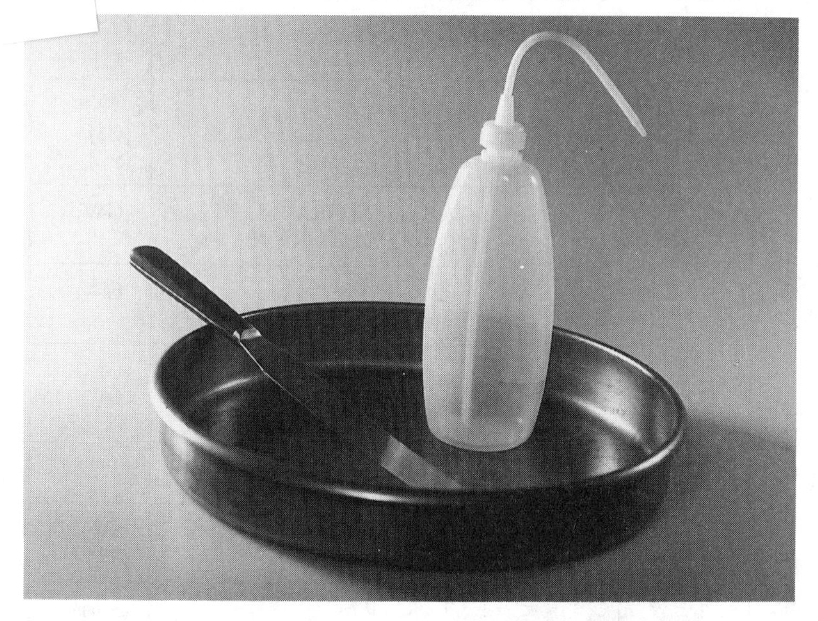

Fig. 4.1 300-mm diameter stainless-steel tray, palette knife and wash bottle.

field classification; this is an important feedback which will improve your field identification of soil types and give you confidence in your judgement.

The items shown in Fig. 4.1 are useful to have when making an examination of soil material on site.

Symbols for soils which can be used in compiling graphical logs etc. are shown in the inside front cover of this handbook.

4.3 Full soil description

Having made a systematic description of both the soil mass and the soil material, the soil can be given a full soil description. It is recommended that the various components of the full soil description always be written in the same order as follows: Compactness or strength, bedding, discontinuities, weathering state, colour, particle shape and composition, soil name. Not all components need feature in every full soil description. The following are three examples: Stiff, very thickly bedded, closely fissured, unweathered, grey CLAY. Dense, medium bedded, yellow, sub-angular, poorly graded SAND. Loose, thickly bedded, slightly weathered, brown, poorly graded GRAVEL with some cobbles. Note that the soil name is written in capitals when it features in the full description.

Table 4.5 Summary of components used in compiling full descriptions of soil. Numbers indicate order in which components are written. Can be used as a checklist on site

1. Compactness or strength
2. Bedding
3. Discontinuities
4. Weathering state
5. Colour
6. Particle shape and composition
7. SOIL NAME

Notes:
1. Not all the components need be used in every full soil description. Choose only those that are appropriate or relevant or that can be observed.
2. Add the geological formation name in brackets after the full description if this is known.
3. Separate the components of the description with commas when writing in a sentence.

When giving the full soil description as described in this chapter, or the full rock description as described in the next chapter, if the official geological formation name is known it can be placed after the full soil or rock description in brackets thus: (Lower Lias Clay) or (Eskdale Granite); this should not be done if there is uncertainty (see Section 8.1).

Table 4.5 summarizes the components used in giving a full soil description and lists the order in which the components should be given; it can be used as a checklist in the field.

4.4 Unified Soil Classification

In the United States, the Unified Soil Classification (USC) system is used to classify engineering soils (Table 4.6) and this is based on laboratory tests. Soils are divided into two major divisions depending upon whether the material is predominantly coarser or finer than 0.075 mm. This separation is between essentially gravelly/sandy soils and essentially silty/clayey soils. Within the major divisions the classification proceeds by finer distinctions down to the soil groups, each of which has a two-letter symbol. The first letter of the symbol refers to the predominant particle size and the second letter of symbol refers either to the grading (for coarse-grained soils) or to the plasticity (for fine-grained soils). The symbols used are listed in Table 4.7.

It is suggested that the system could be used in the field by the engineering geologist if he first becomes familiar with the appearance and feel of samples of soil that have been classified by doing the laboratory tests, but that if used in this way the symbol should be written in brackets, e.g. (SW), to signify that a field rather than a laboratory classification has been made. To make a field identification of soils using the Unified Soil Classification system, the methods of estimating particle size (gravel, sand, silt and clay), grading, plasticity and dilatancy described in Section 4.2 can be used. These can then be applied as shown in Table 4.8 to make an

Table 4.6 Unified Soil Classification system

	Major divisions			*Group symbols*	*Soil names*
GRANULAR SOILS	Coarse-grained soils More than 50 per cent retained on No. 200 sieve*	Gravels 50 per cent or more of coarse fraction retained on No. 4 sieve	Clean Gravels	GW	Well-graded gravels and gravel–sand mixtures, little or no fines.
				GP	Poorly graded gravels and gravel–sand mixtures, little or no fines.
			Gravels with Fines	GM	Silty gravels, gravel–sand–silt mixtures.
				GC	Clayey gravels, gravel–sand–clay mixtures.
		Sands More than 50 per cent of coarse fraction passes No. 4 sieve	Clean Sands	SW	Well-graded sands and gravelly sands, little or no fines.
				SP	Poorly graded sands and gravelly sands, little or no fines.
			Sands with Fines	SM	Silty sands, sand–silt mixtures.
				SC	Clayey sands, sand–clay mixtures.

Table 4.7 Symbols used in the Unified Soil Classification system

Primary group symbols	*Secondary group symbols*
G – Gravel size	W – Well-graded coarse materials
S – Sand size	P – Poorly graded coarse materials
M – Silt size	M – Silt fines
C – Clay size	C – Clay fines
O – Organic material	L – Low plasticity
	H – High plasticity

Table 4.6—continued

	Major divisions		Group symbols	Soil names
COHESIVE SOILS	Fine-grained soils 50 per cent or more passes No. 200 sieve*	Silts and Clays Liquid limit 50 per cent or less	ML	Inorganic silts, very fine sands, rock flour, silty or clayey fine sands.
			CL	Inorganic clays of low to medium plasticity, gravelly clays, sandy clays, silty clays, lean clays.
			OL	Organic silts and organic silty clays of low plasticity.
		Silts and Clays Liquid limit greater than 50 per cent	MH	Inorganic silts, micaceous or diatomaceous fine sands or silts, plastic silts.
			CH	Inorganic clays of high plasticity, fat clays.
			OH	Organic clays of medium to high plasticity.
	Highly Organic Soils		PT	Peat, muck and other highly organic soils.

* Based on the material passing the 75-mm sieve.

U.S. Standard Sieve Size	*Particle size diameter (mm)*
4	4.75
200	0.075

(Adapted from *Annual Book of ASTM Standards*, 1983)

identification of the soil group to which the sample belongs. Where an identification is uncertain, or perhaps on the borderline between two groups, a composite symbol, e.g. (GW-GP), may be used. As well as the properties listed in the table, it is useful to remember that organic soils

Table 4.8 Field identifications of soils of the Unified Soil Classification system

		Group symbols	*Field identification criteria*
GRANULAR SOILS	GRAVELS	(GW)	If the gravel is clean decide if it is well-graded (W) or poorly graded (P) by estimating particle sizes present.
		(GP)	
		(GM)	If the gravel contains fines decide if the fines are silty (M) or clayey (C) by feel.
		GC)	
	SANDS	(SW)	Individual grains of sand can be distinguished by the naked eye. If the sand is clean proceed as above (W or P). If the sand contains fines proceed as above (M or C).
		(SP)	
		(SM)	
		(SC)	
COHESIVE SOILS	SILTS AND CLAYS	(ML)	If the soil is of low plasticity decide if it is silt (M), clay (C) or organic (O) on the basis of dilatancy (silt) and colour (organic soils will be very dark).
		(CL)	
		(OL)	
		(MH)	If the soil is of high plasticity decide if it is silt (M), clay (C) or organic (O) as above.
		(CH)	
		(OH)	
		(PT)	Contains recognizable fibrous material. Dark.

sometimes have a characteristic smell of decayed vegetation. A complete description of the field method of using the USC system is given in *ASTM: D2488-84* (American Society for Testing and Materials, 1988).

When reporting a soil identification using the Unified Soil Classification as described above, or when reporting a rock identification as described in Section 5.4, if the official geological formation name is known, it can be given as well, e.g. Kansan Till or Diamond Creek Sandstone.

5
Description of rocks

A complete description of rocks comprises two parts: a description of the rock mass characteristics and a description of the rock material characteristics. The description of the rock material leads to the *rock name*, and this together with the description of the rock mass leads to the *full rock description*. It is recognized that to make a full description of rocks in the field, and in particular to identify rocks correctly, is something that can only be done properly after geological training, and it is therefore assumed here that the reader will have had some geological field instruction. In this context, the three Handbooks *The Field Description of Sedimentary Rocks, The Field Description of Metamorphic Rocks* and *The Field Description of Igneous Rocks* in this series are valuable guides. In this book a basic geological knowledge of rocks will be assumed and attention will be concentrated on the engineering aspects of rock description.

However, for those readers without basic geological training, the following generalized description of igneous, sedimentary and metamorphic rocks is given. It is assumed that a hand specimen is being examined and that the penknife, hand lens and acid bottle (see Table 2.1) are being used. Note that fine-grained, weathered or altered rocks are difficult or impossible to identify with certainty by field methods alone. Following this introduction we turn to the main topic of the field description of engineering rocks.

- *Igneous rocks* are those that have formed from the crystallization on cooling of molten material (magma), either in the earth's crust or erupted at the surface. Coarse-grained and medium-grained igneous rocks have crystals in them that are visible to the naked eye or with the hand lens; these crystals cannot be detached from the rock with the point of a penknife blade. Fine-grained igneous rocks cannot be scratched with the penknife (or scratched only with difficulty). Igneous rocks are usually strong and hard.
- *Sedimentary rocks* are formed from fragments, grains and minerals derived from pre-existing rocks by the action of denudation, transport and deposition. They are

often layered in appearance due to bedding, caused by variations in composition of the arriving sediment. The grains of sedimentary rocks can usually be detached with the point of a penknife if they are coarse-grained or medium-grained, and fine-grained sedimentary rocks can be scratched with the penknife. Indurated sedimentary rocks can be strong and hard, but non-indurated ones may be weak and soft. Limestones and chalk effervesce on treatment with acid.

- *Metamorphic rocks* are formed by the action of heat or pressure (or both acting together) on pre-existing rocks. In coarse-grained or medium-grained varieties crystals may be visible to the naked eye or with the hand lens. Metamorphic rocks are sometimes banded or foliated due to segregation or recrystallization of minerals during the metamorphic process. Metamorphic rocks are usually strong and hard. Marble effervesces on treatment with acid.

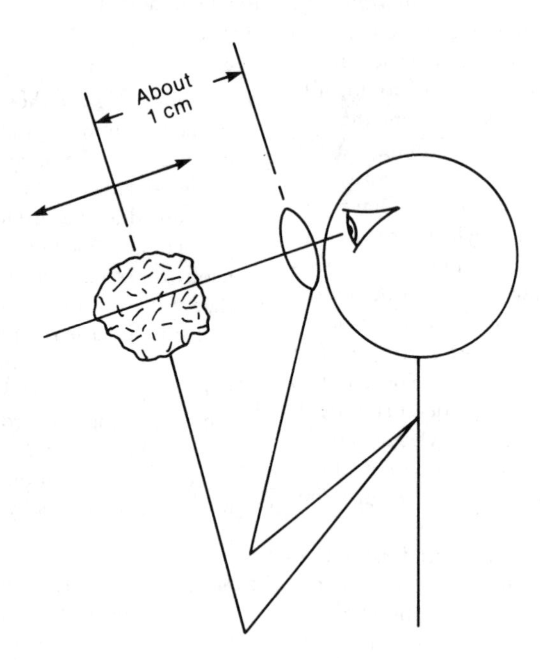

Fig. 5.1 Correct way to use hand lens: hold lens close to eye and move rock specimen to-and-fro until in focus.

5.1 Description of rock material

Description of rock material is carried out on a hand specimen of rock obtained from a natural exposure such as a cliff or an outcrop, an artificial exposure such as a quarry or trial pit, or from a borehole core. Using a geological hammer (and chisel if necessary) a *fresh* surface of rock is prepared for examination. Using a hand lens (Fig. 5.1), the grain size, texture and fabric of the rock are then examined. A scale of grain size, which refers to the average size of the minerals or particles comprising the rock, is given in Tables 5.1, 5.2 and 5.3. The texture is the general physical appearance or character of a rock arising from the mutual relationship of the components, and rock texture may be, for example, granular, crystalline, amorphous, porphyritic. The fabric refers to the arrangement of the components and rock fabric may be, for example, bedded, foliated, cemented. Using the reaction to hydrochloric acid, limestone may be distinguished from magnesian limestone or dolomite. Granular sedimentary rocks can be distinguished from igneous and metamorphic rocks by scratching with a penknife which should remove whole grains from the former.

Having observed these characteristics of the specimen it should be possible to select a rock name from those given in bold capital letters in Tables 5.1, 5.2 and 5.3. Do not be afraid of giving a provisional name if identification is difficult; there is nothing wrong in describing a rock as a 'dark grey, medium-grained, crystalline rock' and leaving the exact rock name until you get back to the laboratory and can make a more thorough examination. Also, if petrographical analyses are later made of any specimens you collect, use these to check on your field identification; in this way you can build up experience and gain confidence. Symbols for rocks which can be used in compiling graphical logs, etc. are shown in the inside back cover of this handbook.

The colour of rocks can be described using the simple names given previously for soils or, more fully, using the *Rock Color Chart* published by the Geological Society of America (1963). The strength of rocks can be assessed using the geological hammer (see Table 6.1), or measured using the simple tests described in Chapter 6. Although the colour and strength are not required to give the rock name, they are required for the full rock description to be described shortly.

5.2 Description of rock mass

Description of rock mass is made by standing back and observing the field setting of the rock in its whole exposure, be it outcrop, cliff, quarry or cutting. In particular three things should be noted: structure, discontinuities and weathering. The structure of sedimentary rocks included bedding, dip, folding, interbedding,

Table 5.1 Field identification of sedimentary rocks

Grain size (mm)	*Bedded rocks*						
More than 20	Grain size description			At least 50 per cent of grains are of carbonate		At least 50 per cent of grains are of fine-grained volcanic rock	Chemical and carbonaceous rocks
20 – 6 –	RUDACEOUS		**CONGLOMERATE** Rounded boulders, cobbles and gravel cemented in a finer matrix **BRECCIA** Irregular rock fragments in a finer matrix	STONE and DOLOMITE 2.1)		Fragments of volcanic ejecta in a finer matrix Rounded grains **AGGLOMERATE** Angular grains **VOLCANIC BRECCIA**	**HALITE** **ANHYDRITE**
2 –	CEOUS	Coarse	**SANDSTONE** Angular or rounded grains, commonly cemented by clay, calcitic or iron minerals		**OOLITIC LIMESTONE**	Cemented volcanic ash	**GYPSUM**

0.6 – 0.2	ARENA	Medium	**SEDIMENTARY QUARTZITE** Quartz grains and siliceous cement		**LIMESTONE, MAGNESIAN LIME** (see Table		**TUFF**	
0.2 – 0.06		Fine	**ARKOSE** Many feldspar grains **GREYWACKE** Many rock chips					
0.06 – 0.002	ARGILLACEOUS		**MUD-STONE**	**SILT-STONE** Mostly silt		**CHALK**	**FINE-GRAINED TUFF**	
0.002 – Less than 0.002			**SHALE** Fissile	**CLAY-STONE** Mostly clay			**VERY FINE-GRAINED TUFF**	
Amorphous or crypto-crystalline			**FLINT:** occurs as bands of nodules in the Chalk **CHERT:** occurs as nodules and beds in limestone and calcareous sandstone					**COAL LIGNITE**

(Adapted from BS 5930:1981)

Table 5.2 Field identification of metamorphic rocks

Foliated rocks		*Non-foliated rocks*	*Grain size (mm)*
Grain size description			More than 20
COARSE	**GNEISS** Well developed but often widely spaced foliation sometimes with schistose bands **MIGMATITE** Irregularly foliated: mixed schists and gneisses	**MARBLE** **METAMORPHIC QUARTZITE** **GRANULITE** **HORNFELS** **AMPHIBOLITE** **SERPENTINE**	20 6 2
MEDIUM	**SCHIST** Well developed undulose foliation; generally much mica		0.6 0.2 0.06
FINE	**PHYLLITE** Slightly undulose foliation; sometimes 'spotted' **SLATE** Well developed plane cleavage (foliation)		0.002 Less than 0.002
	MYLONITE Found in fault zones, mainly in igneous and metamorphic areas		Amorphous or cryptocrystalline

(Adapted from BS 5930:1981)

false bedding etc., that of metamorphic rocks includes foliation, banding and cleavage, while igneous rocks may be intrusive or extrusive and may show a massive structure or be flow-banded. Special structures such as columnar jointing should be noted. Descriptive terms for the spacing of more-or-less planar structural features in rocks are given in Table 5.4. Examples of the use of these terms are: thinly bedded sandstone, medium foliated gneiss, very thickly flow-banded diorite.

All breaks in the rock including bedding plane partings, joints, faults, cracks and other fractures are grouped together by the engineering geologist under the term *discontinuities*. The spacing, orientation and persistence of discontinuities should be recorded together with notes as to whether they are open, closed, cemented or filled. The surface of discontinuities may be planar, curved, irregular, slickensided, smooth or rough and these characteristics should be noted as well. Probably the best way of recording discontinuities is to set up a *scanline* on the exposed rock face to be examined. A surveyor's measuring tape is useful for this purpose and it can be attached to the rock by hammering 6-inch nails into convenient joints and suspending the tape from them. The length of the scanline will depend on the site, but a length of 10 m is suggested for a start. The orientation of the scanline has to be chosen with regard to the rock structure and more than one may be necessary to adequately record the discontinuities. In sedimentary rocks, for example, the minimum will be one scanline along the bedding direction and one perpendicular to it. The compass-clinometer is used to measure the orientation of individual discontinuities which is conventionally recorded as two numbers separated by an oblique line, the dip angle being given first followed by the dip direction which is *always* given as a three digit number. For example, 60/090 defines a discontinuity dipping at 60° to the east. The openness of a discontinuity, called its *aperture* is most easily assessed with a simple home-made feeler gauge (Fig. 2.1) and can be described using a matching scale such as: closed, less than 0.1 mm; tight, 0.1 to 1 mm; open, 1 to 5 mm; and wide, greater than 5 mm. Scales for describing the spacing of bedding and discontinuities are given in Table 4.1. Some criterion will need to be chosen to exclude very small discontinuities which may be of little significance from an engineering point of view; it is suggested that those less than 0.15 m long could be disregarded to begin with, but this may need to be revised in the light of the particular application.

Before leaving the subject of discontinuity measurement, mention should be made of Rock Quality Designation (RQD). This is a system of describing the fracture state of rock core, the logging of which is generally outside the scope of this book. To measure the RQD, the *natural* fractures in the core and not the ones produced by drilling and handling are

Table 5.3 Field identification of igneous rocks

Rocks with massive structure and crystalline texture					*Grain size (mm)*
Grain size description	PEGMATITE				More than 20
COARSE	GRANITE	DIORITE	GABBRO	PYROXENITE PERIDOTITE	20 6
	These rocks are sometimes porphyritic and are then described, for example, as porphyritic granite				2
MEDIUM	MICRO-GRANITE	MICRO-DIORITE	DOLERITE		0.6
	These rocks are sometimes porphyritic and are then described as porphyries				0.2

FINE	**RHYOLITE** These rocks are sometimes porphyritic and are then described as porphyries	**ANDESITE**	**BASALT**		0.06 – 0.002
	OBSIDIAN	**VOLCANIC GLASS**			Less than 0.002 Amorphous or cryptocrystalline
	Pale ← colour → Dark				
	ACID Much quartz	INTERMEDIATE Some quartz	BASIC Little or no quartz	ULTRA BASIC	

(Adapted from BS 5930:1981)

Table 5.4 Descriptive terms used for the spacing of planar structures in rocks

Term	*Spacing (mm)*
Very thick	greater than 2000
Thick	600 to 2000
Medium	200 to 600
Thin	60 to 200
Very thin	20 to 60
Thickly laminated (sedimentary) Narrow (metamorphic and igneous)	6 to 20
Thinly laminated (sedimentary) Very narrow (metamorphic and igneous)	less than 6

(After BS 5930:1981)

first identified. The RQD is the percentage of rock core recovered in intact lengths that are 100 mm or more long, only core lengths terminated by natural fractures being considered. Measurement should be made along the core axis. Be particularly wary or rock core that consists of a large number of discs – these can be produced by bad drilling. Having determined the RQD, the descriptive terms given in Table 5.5 may be applied to the rock core. Note that RQD, strictly speaking, can be applied to rock core only and not to other means of sampling rock such as rock exposures in natural outcrops or in quarries, although it is sometimes used in these other ways.

Table 5.5 Descriptive terms for RQD values

RQD (per cent)	*Term*
0 to 25	Very poor
25 to 50	Poor
50 to 75	Fair
75 to 90	Good
90 to 100	Excellent

(After BS 5930:1981)

The weathering state of the rock mass is described and it should be noted that weathering may proceed through the rock material itself which will therefore be changed, or along the surfaces of discontinuities, leaving blocks or 'corestones' relatively unchanged. Table 5.6 gives a description of what to look for, together with terms and a grade scale appropriate to each degree of weathering. Table 5.6 is applicable to all rocks except chalk, which by traditional usage has a weathering grade scale of its own given in Table 5.7. When working in tropical countries it is very important to realize that weathering may have

Table 5.6 Scale of weathering grades of rock mass

Term	*Description*	*Grade*
Fresh	No sign of rock material weathering; perhaps slight discoloration on major discontinuity surfaces.	I
Slightly weathered	Discoloration indicates weathering of rock material and discontinuity surfaces. All the rock material may be discoloured.	II
Moderately weathered	Less than half of the rock material is decomposed or disintegrated to a soil. Fresh or discoloured rock is present either as a continuous framework or as corestones.	III
Highly weathered	More than half of the rock material is decomposed or disintegrated to a soil. Fresh or discoloured rock is present either as a discontinuous framework or as corestones.	IV
Completely weathered	All rock material is decomposed and/or disintegrated to soil. The original mass structure is still largely intact.	V
Residual soil	All rock material is converted to soil. The mass structure and material fabric are destroyed. The soil has not been significantly transported.	VI

(After BS 5930:1981)

proceeded to a great depth and there may be *no* exposures of fresh rock either at the surface or within the normal depths of civil engineering operations.

5.3 Full rock descriptions

Having described the rock material (and given the rock a rock name as a consequence) and the rock mass, they can now be brought together in a full rock description. As for soil description it is customary always to write the components in the same order: Colour, grain size, structure, weathering state, rock name, strength, discontinuities. If they have not all been determined or if they are not all

Table 5.7 Weathering grades of chalk

Grade	*Description*
VI	Extremely soft structureless chalk containing only small lumps of intact chalk.
V	Structureless chalk containing lumps of intact chalk.
IV	Rubbly partly weathered chalk with bedding and jointing. Joints 10–60 mm apart, open to 20 mm and often infilled with soft remoulded chalk and fragments.
III	Rubbly to blocky unweathered chalk. Joints 60–200 mm apart, open to 3 mm and sometimes infilled with fragments.
II	Blocky medium-hard chalk. Joints more than 200 mm apart and closed.
I	As for Grade II but hard and brittle.

(After Wakeling, 1969)

appropriate, not all components need feature in the full description. Examples of full rock descriptions for a sedimentary rock, a metamorphic rock and an igneous rock respectively are: Red, coarse grained, thickly bedded, slightly weathered SANDSTONE, moderately weak, with tight widely spaced joints 80/270. Greenish grey, medium grained, thinly foliated, fresh GNEISS, strong. Pinkish grey, coarse grained and porphyritic, massive, fresh GRANITE, extremely strong, with two sets of closed, very widely spaced joints 85/355 and 75/095. Note that the rock name is written in capitals when it features in the full description.

Table 5.8 summarizes the components used in giving a full rock description and lists the order in which the components should be given; it can be used as a checklist in the field.

Table 5.8 Summary of components used in compiling full descriptions of rock. Numbers indicate order in which components are written. Can be used as a checklist on site

1. Colour
2. Grain size
3. Structure
4. Weathering state
5. ROCK NAME
6. Strength
7. Discontinuities

Notes:

1. Not all the components need be used in every full rock description. Choose only those that are appropriate or relevant or that can be observed.
2. Add the geological formation name in brackets after the full description if this is known.
3. Separate the components of the description with commas when writing in a sentence.

Table 5.9 US rock classification

Class	*Type*	*Family*
Igneous	Intrusive (coarse-grained)	Granite Syenite Diorite Gabbro Peridotite Pyroxenite Hornblendite
	Extrusive (fine-grained)	Obsidian Pumice Tuff Rhyolite Trachyte Andesite Basalt Diabase
Sedimentary	Calcareous	Limestone Dolomite
	Siliceous	Shale Sandstone Chert Conglomerate Breccia
Metamorphic	Foliated	Gneiss Schist Amphibolite Slate
	Nonfoliated	Quartzite Marble Serpentinite

(After US Department of Commerce)

5.4 US rock classification

Engineering geologists working in the United States can follow the US Department of Commerce system for the general classification of rock (Table 5.9). The field identification of the rock types is carried out by examination of a freshly broken surface of a hand specimen of rock. A hand lens, a good quality steel penknife and a dropping bottle containing dilute hydrochloric acid are also required to examine the mineralogical composition and texture, to test for hardness and to test for effervescence. The fully worked out scheme of examination is given in Table 5.10, in which the rock names are shown in italics.

Using the classification system, an attempt should be first made to place the rock in the appropriate group or subgroup, and having done this to try and proceed to an identification of the individual rock type. Experience has shown that difficulty is sometimes found in identifying the intrusive igneous rocks, and the engineering geologist when beginning to use the classification should try and be familiar with the appearance of hand specimens of rocks that have already been classified in the laboratory.

5.4.1 Unified Rock Classification system

Having identified a rock using the general geological classification described in the foregoing section, an engineering classification should then be made, and in the United States the Unified Rock Classification (URC) system is widely used to do this. The system utilizes four elements of the rock's properties, and all can be estimated or measured in the field using the following simple equipment: a ×10 hand lens, a 1 lb ball-peen hammer, and a spring-balance together with a bucket of water.

The basic elements in the classification system are:

1 *The degree of weathering*, estimated by the freshness of the rock as observed by eye or with the hand lens.
2 *The strength*, estimated by the reaction of the rock to a blow struck with the standard hammer.
3 *The planar and linear elements*, estimated by observing the discontinuities of the rock mass.
4 *The unit weight*, measured by weighing a specimen of the rock in air and in water using the spring-balance and bucket.

Each element in the classification system is given a rating on a five-point scale, ABCDE, in which A is best and E is worst. This rating is known as the Category Symbol and the criteria for each are shown in Table 5.11.

When carrying out tests in the field, the standard abbreviations given in Table 5.11 can be used to record your observations concisely in your field notebook. When estimating the strength, hit the rock *hard* with the rounded end of the hammer head, examine the impact site and make an estimation of the rock strength guided

Table 5.10 Field identification for US rock classification

SYSTEM FOR THE IDENTIFICATION OF COMMON ROCKS.
PRELIMINARY CLASSIFICATION

Group I. – Glassy, wholly or partly.
Group II. – Not glassy; dull or stony; homogeneous; so fine-grained that grains cannot be recognized.
Group III. – Distinctly granular.
Group IV. – Distinctly foliated; no effervescence with acid.
Group V. – Clearly fragmental in composition; rounded or angular pieces or grains cemented together.

GROUP I. – GLASSY ROCKS

1. Glassy luster; hard; conchoidal fracture; colorless to white or smoky gray; generally brittle. *Quartz*.
2. Solid glass; may have spherical inclusions; brilliant vitreous luster; generally black. *Obsidian*.
3. Cellular or frothy glass. *Pumice*.

GROUP II. – DULL OR STONY, VERY FINE-GRAINED ROCKS

SUBGROUP II A. – Not scratched by fingernail, but readily scratched with knife.

1. Particles almost imperceptible; dull luster; homogeneous; clay odor; little if any effervescence with acid; laminated structure; breaks into flakes. *Shale*.
2. Little if any clay odor; brisk effervescence with acid. *Limestone*.
3. Little if any clay odor; brisk effervescence with acid only when rock is powdered or acid is heated. *Dolomite*.
4. Soapy or greasy feel; translucent on thin edges; green to black; no effervescence. *Serpentinite*.

SUBGROUP II B. – Not scratched with the knife or scratched only with difficulty; no effervescence with acid.

1. Light to gray color; clay odor possible; may have a banded flow structure. *Felsite*.
2. Very hard; pale colors to black; no clay odor; conchoidal fracture; waxy or horny appearance. *Chert*.
3. Heavy; dark color; may have cellular structure; may contain small cavities filled with crystalline minerals. *Basalt*.

continued

Table 5.10—continued

GROUP III. – GRANULAR ROCKS

SUBGROUP III A. – Easily scratched with the knife.

1. Brisk effervescence with acid. *Limestone* or *Marble*.
2. Brisk effervescence only with warm acid, or with powdered rock. *Dolomitic marble*.

SUBGROUP III B. – Hard; not scratched with knife or scratched with difficulty; grains of approximately equal size.

1. Mainly quartz and feldspar; usually light colored, sometimes pinkish. *Granite*.
2. Mainly feldspar; little quartz (less than 5 per cent); light colors of nearly white to light gray or pink. *Syenite*.
3. Feldspar and a dark ferromagnesian mineral.
   - (a) Major constituent feldspar; rock of medium color. *Diorite*.
   - (b) Ferromagnesian mineral equal to or in excess of feldspar; rock of dark color.
     - (1) Grains just large enough to be recognized by the unaided eye. *Diabase*.
     - (2) Coarse-grained rock. *Gabbro*.
4. Mainly ferromagnesian minerals; generally dark green to black.
   - (a) Predominant olivine with pyroxene or hornblende. *Peridotite*.
   - (b) Predominant augite. *Pyroxenite*.
   - (c) Predominant hornblende. *Hornblendite*.
5. Mainly quartz.
   - (a) Fracture around grains. *Sandstone*.
   - (b) Fracture through all or through an appreciable percentage of grains. *Quartzite*.

SUBGROUP III C. – Hard; not scratched with knife or scratched with difficulty; large distinct crystals in finer groundmass.

1. Crystals of feldspar and quartz with some of a ferromagnesian mineral (generally biotite) in a light-colored groundmass. *Granite porphyry*.
2. Crystals of feldspar and usually a ferromagnesian mineral in a light-colored groundmass. *Syenite porphyry*.
3. Crystals of ferromagnesian minerals, or of striated feldspar, or both, in a medium-colored groundmass. *Diorite porphyry*.
4. Crystals of quartz, or feldspar, or both, generally with a ferromagnesian mineral, in a predominant, fine-grained groundmass of light color. *Felsite porphyry*.
5. Crystals of feldspar, or of a ferromagnesian mineral, or both, in a fine-grained, dark or black, heavy groundmass. *Basalt porphyry*.

GROUP IV. – FOLIATED ROCKS

1. Medium to coarse grain; roughly foliated. *Gneiss.*
2. More finely grained and foliated. *Schist.*
   - (a) Consists mainly or largely of mica with some quartz. *Mica schist.*
   - (b) Medium green to black; consists mostly of a felted or matted mass of small, bladed or needle-like crystals arranged in one general direction. *Hornblende schist* or *amphibolite.*
   - (c) Glassy or silky luster on foliation surfaces; splits readily into thin pieces. *Sericite schist.*
   - (d) Soft, greasy feel; marks cloth; easily scratched with fingernail; whitish to light gray, or green. *Talc schist.*
   - (e) Smooth feel; soft; glimmering luster; green to dark green. *Chlorite schist.*
3. Very fine grain; splits easily into thin slabs; usually dark gray, green or black. *Slate.*

GROUP V. – FRAGMENTAL

1. Rounded pebbles embedded in some type of a cementing medium. *Conglomerate.*
2. Angular fragments embedded in a cementing medium. *Breccia.*
3. Fragments of volcanic (fine-grained or glassy) rocks embedded in compacted volcanic ash. *Volcanic tuff* or *Volcanic breccia.*
4. Quartz grains, rounded or angular, cemented together. *Sandstone.*
5. Quartz and feldspar grains cemented together to resemble the appearance of granite. *Arkose* (feldspathic sandstone).

(After US Department of Commerce)

Table 5.11 Unified Rock Classification system

Category symbol	*Abbreviation*	*Observation*
Degree of Weathering Element		
A	MFS	Micro Fresh State (determined by hand lens)
B	VFS	Visual Fresh State
C	STS	Stained State
D	PDS	Partly Decomposed State
E	CDS	Completely Decomposed State
Strength Estimate Element		
A	RQ	Rebound reaction to hammer blow
B	PQ	Pits with hammer blow
C	DQ	Dents with hammer blow
D	CQ	Craters with hammer blow
E	MBL	Can be remolded with finger pressure
Planar and Linear Elements		
A	SRB	Solid with random breakage
B	SPB	Solid with preferred breakage
C	LPS	Latent planes of separation
D	2-D	Planes of separation in two dimensions
E	3-D	Planes of separation in three dimensions
Unit Weight Element		
A	> 160	Greater than 160 lb/ft^3 (> 2.55 Mg/m^3)
B	150–160	150–160 lb/ft^3 (2.40–2.55 Mg/m^3)
C	140–150	140–150 lb/ft^3 (2.25–2.40 Mg/m^3)
D	130–140	130–140 lb/ft^3 (2.10–2.25 Mg/m^3)
E	< 130	Less than 130 lb/ft^3 (< 2.10 Mg/m^3)

(After Williamson and Kuhn, 1988)

by Fig. 5.2. Wear safety goggles when carrying out this strength test. Category E material in the strength element of the classification can also be examined as an engineering soil rather than a rock using the Unified Soil Classification system described in Section 4.4; the same applies to Categories D and E in the degree of weathering element.

Category A: Rebound Quality (RQ)–

Category B: Pit Quality (PQ)–

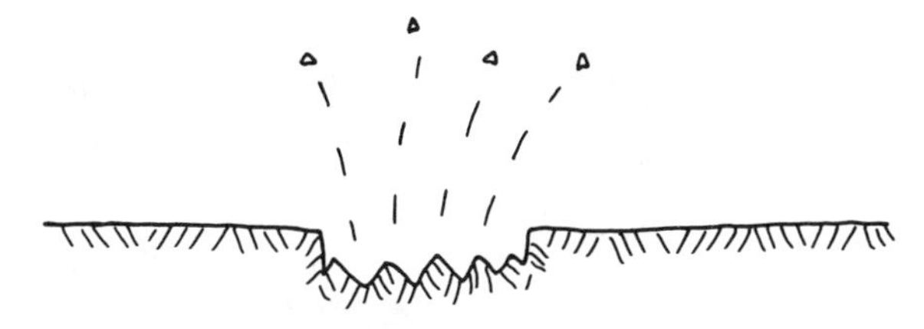

Category C: Dent Quality (DQ)–

Category D: Crater Quality (CQ)–

Fig. 5.2 Reaction of rock to ball-peen hammer impact. (After Williamson and Kuhn, 1988.)

The unit weight is measured in the field by attaching a piece of clean intact rock to the spring balance (one with a 0–10 lb or 0–5 kg range is suitable) with a piece of string (Fig. 5.3). The weight of the specimen is noted first in air and then when submerged in water, ignoring the weight of the string. The unit weight of the rock is given by:

$$\text{Unit weight} = \frac{W_a}{W_a - W_w} \times D_w$$

where W_a is the weight in air, W_w is

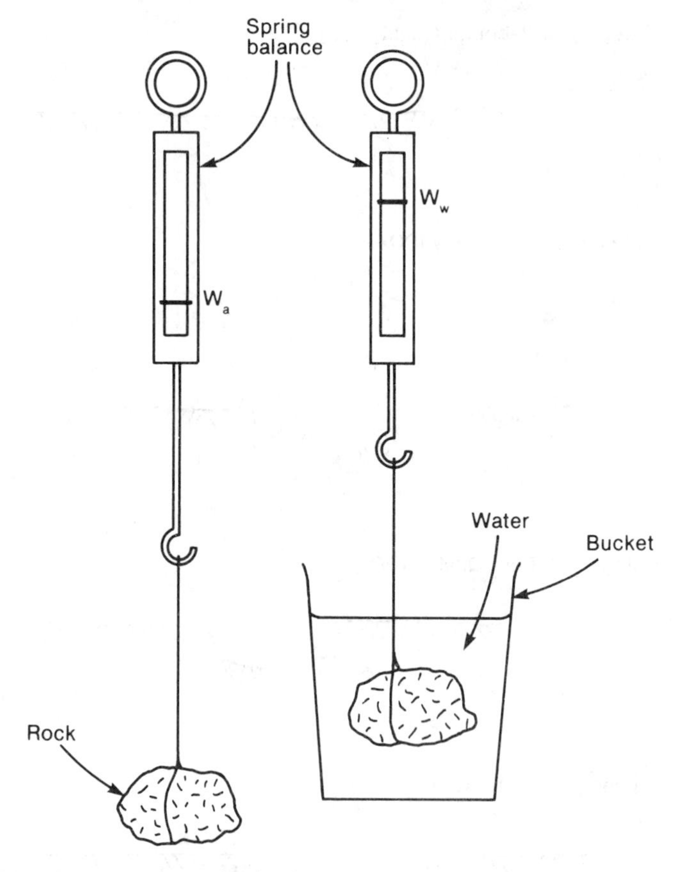

Fig. 5.3 Measuring unit weight of rock.

the weight in water, and D_w is the unit weight of water (62.4 lb/ft^3 or 1.00 Mg/m^3). It may not be possible to determine the unit weight of Category D and E material in the degree of weathering element if the specimen disintegrates on immersion in water.

The degree of weathering element is estimated by visual examination of the state of oxidation (brown coloration), alteration, and decomposition of the rock, using both the naked eye and the hand lens. The planar and linear elements are estimated by assessing the absence or presence of preferred-breakage direction, cleavage, and two-dimensional or three-dimensional discontinuities.

When formally reporting the results of a Unified Rock Classification rating, both the category symbol and the abbreviation should be given for each element, and they should be written as a fraction with the category symbol on top, for example $\frac{\text{B}}{\text{VFS}}$.

The elements should be given in the order shown in Table 5.11 thus:

$$\frac{\text{B}}{\text{VFS}}\ \frac{\text{C}}{\text{DQ}}\ \frac{\text{A}}{\text{SRB}}\ \frac{\text{D}}{\text{130–140}}$$

This signifies a visually fresh, moderately strong, solid, low unit weight, rock. A complete description of the URC system is given by Williamson and Kuhn (1988).

6
Measurement of rock and soil strength

An estimate of the strength of rocks and soils can be made using the simple qualitative methods described in Tables 4.2 and 6.1 but whenever possible the engineering geologist is urged to supplement this by quantitative measurement. Simple test apparatus is now available for carrying out rock and soil strength measurements in the field and these will now be described. Detailed instructions for using each piece of equipment is provided by the suppliers. With each of the tests to be described, for each rock or soil tested, repeat determinations (say 10 to 20) should be made and the mean value taken. The calibration of the apparatus should also be checked periodically. The apparatus should be kept clean and great care taken to keep rock dust or soil from the moving parts, scales, etc.

When recording the results of tests always be sure to state the units of measurement – some instruments have multiple scales and you may not be able to remember afterwards which one you used. Where the apparatus is supplied in a case, put it back in the case when testing is finished and when doing so check that none of the parts or accessories is missing.

6.1 Field rock strength tests

The *Schmidt hammer* (Fig. 6.1) is a portable instrument for measuring the rebound from the surface of a material. The Schmidt hammer is used by pressing the plunger against the surface to be tested. This causes a spring-loaded mass to be released which strikes the plunger and rebounds, moving an index pointer up a scale numbered from 10 to 100 which indicates the rebound number R. For testing rock surfaces, the types N or L Schmidt hammer should be used. The rebound number can be converted to unconfined compressive strength by carrying out a series of comparative tests and constructing a correlation curve; Fig. 6.2 shows such a curve for a type N hammer and a range of Carboniferous rocks. The Schmidt hammer will be mainly used for testing *in situ* rock surfaces in natural exposures, quarries, tunnel walls, rock cuttings, etc. but it can also be

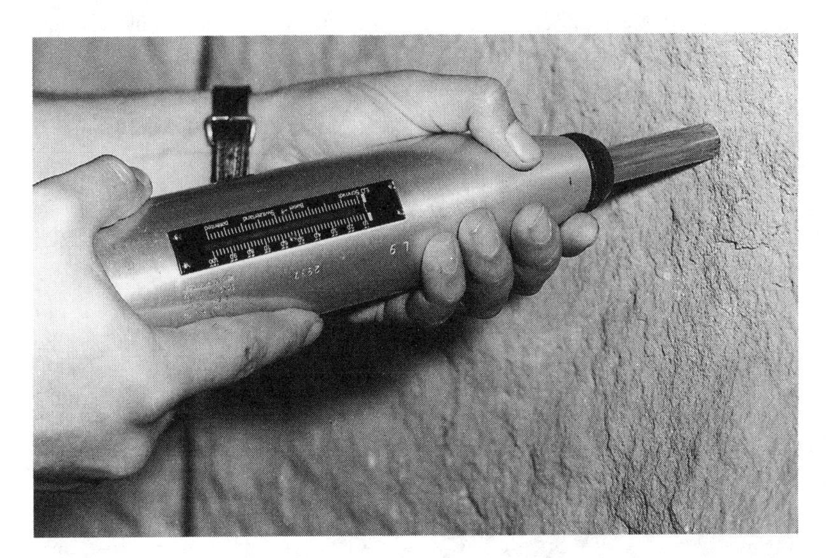

Fig. 6.1 Schmidt hammer.

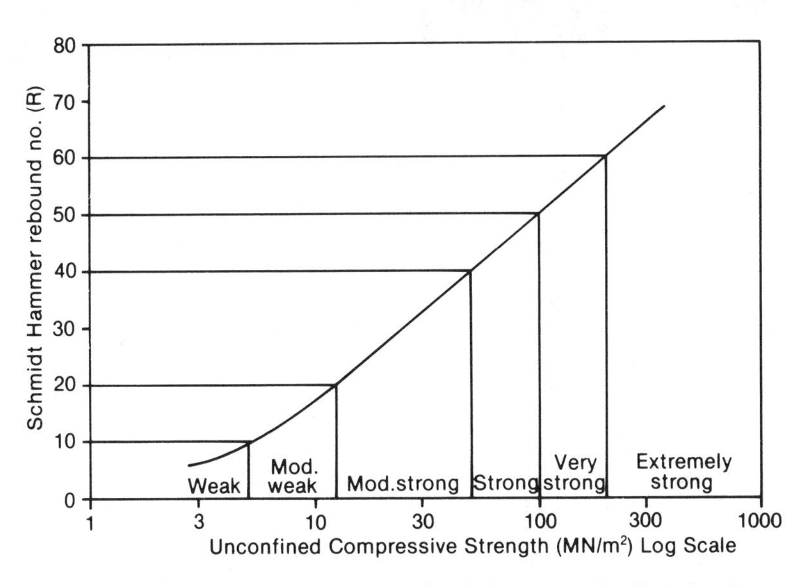

Fig. 6.2 Correlation curve for Type N19 Schmidt hammer used with Carboniferous rocks. (After Carter and Mills, 1976.)

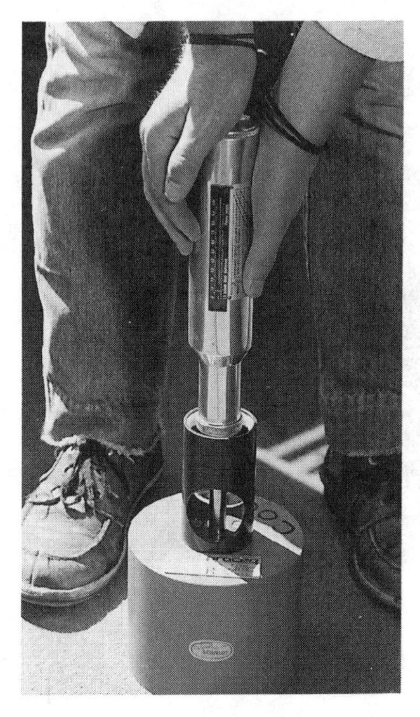

Fig. 6.3 Checking Schmidt hammer with special testing anvil.

used for testing the surface of large diameter rock core so long as a proper support for the rock core is provided. The Schmidt hammer should be checked at site temperature before and after each period of use with the special testing anvil supplied by the manufacturer as shown in Fig. 6.3.

When tested on the anvil, the type N hammer should give a reading of 79 ± 2, and the type L hammer should give a reading of 73 ± 2 (or the values stamped on the anvil). If the rebound number is within these ranges no further action need be taken. If the rebound number is outside these ranges then a correction must be applied as follows:

$$R = R_r \times \frac{79}{R_a} \text{ for the type N hammer,}$$

or

$$R = R_r \times \frac{73}{R_a} \text{ for the type L hammer,}$$

where R_r is the rebound value obtained on the rock and R_a is the rebound value obtained on the anvil. If the rebound number on the anvil is less than 72 for the type N hammer, or 66 for the type L hammer, the instrument should be dismantled, cleaned and recalibrated.

When testing an *in situ* rock surface, care should be taken that the surface is clean and relatively flat. If a hollow sound is produced when making a test it indicates that the rock surface may be separated from the rock mass by a joint and another test position should be selected. It should be noted that the Schmidt hammer has a light-alloy housing and, therefore, cannot be taken down coal mines unless a brass housing is specially made for it.

The *point load tester* (Fig. 6.4) is a field apparatus for measuring the crushing strength of a hand sample of rock. The rock specimen, about 5 cm in size and roughly cubical, is placed between the conical test platens of the apparatus and a load is applied by a hand-operated hydraulic pump. The hydraulic pressure at failure is indicated on a pressure gauge which carries an index pointer to show the

Fig. 6.4 Point load tester.

maximum reading. The failure load P is obtained by multiplying the pressure gauge reading by the cross-sectional area of the ram, the value of which is engraved on the apparatus. The point load strength I_s is given by P/D^2 where D is the distance between the platens at the end of the test and corresponds to the thickness of rock tested; the value of D can be read off the apparatus on a scale provided. Three pressure gauges are supplied to cover different ranges of rock strength; they are attached to the apparatus with a quick-fit hydraulic connector. Provided that the specimen thickness is fairly close to 50 mm, the unconfined compressive strength of the rock is given by multiplying I_s by 23.7. (If the specimen is much larger or smaller, a size correction must be made first. For weak rocks you are advised to determine a site-specific correlation factor yourself.) When testing sedimentary rocks, half the number of specimens should be tested perpendicular to the bedding and half parallel to the bedding; the same applies to other rocks having a marked anisotropy, such as schist, gneiss or slate. Although the point load tester can be used to test rock core, the great advantage of the apparatus is that it can be used to test irregular lumps of rock, thus doing away with the need for elaborate specimen preparation. Although it is not practicable to carry the point load tester very far by hand it can, of course, be easily transported by car to a convenient location on site. Further details of the test are given by Broch and Franklin (1972).

The *NCB [National Coal Board, now British Coal] cone indenter* (Fig. 6.5) is a small portable test apparatus which measures the penetration of a tungsten carbide conical tipped indenter into a rock specimen under a standard applied load. The specimen, a chip of rock not larger than $12 \times 12 \times 6$ mm, is tested by being placed on the flat steel strip and the indenter is then applied to the rock surface by turning the micrometer screw. The standard cone indenter number I_s, or the modified cone indenter number I_m, are calculated from a combination of the micrometer and the dial gauge readings. The procedure, readings and calculations are all set out in a handbook which comes with the apparatus (National Coal Board, 1972). The unconfined compressive strength of the rock in MN/m^2 can be obtained from the cone indenter numbers by multiplying I_s by 24.8, or I_m by 35.8. The cone indenter test should be limited to rocks with a grain size of less than about 0.05 mm

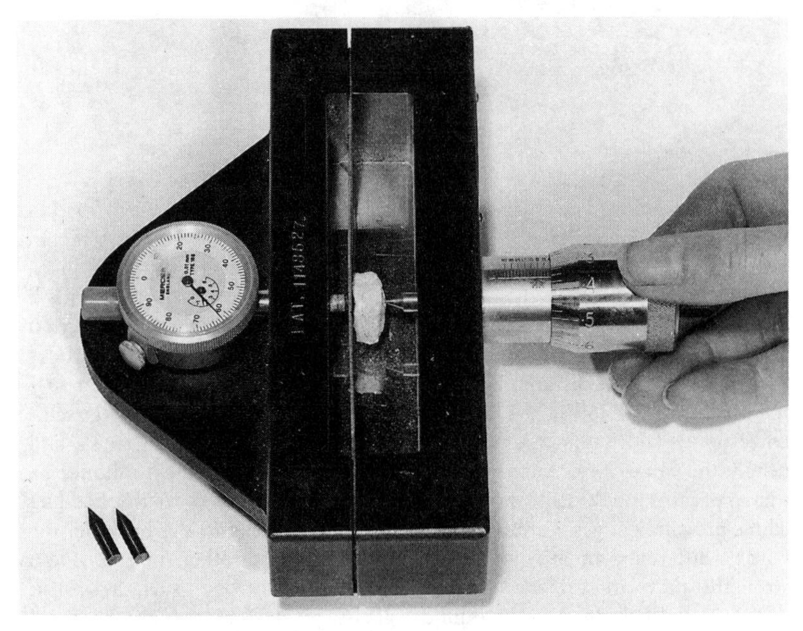

Fig. 6.5 NCB cone indenter. Spare points are shown on the left.

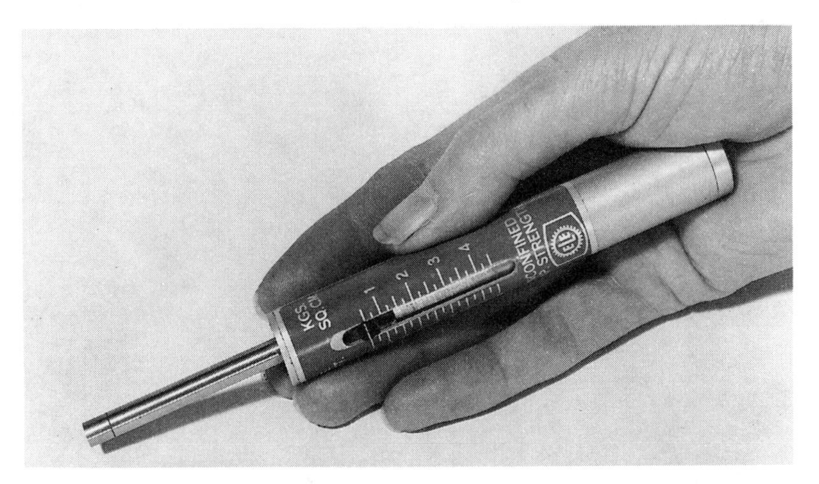

Fig. 6.6 Pocket penetrometer.

because with coarser grained rocks there is a danger of testing individual grains rather than the rock itself. When the conical tip becomes worn it must be changed; replacements can be obtained from the manufacturer. Before using the instrument, check with a colleague that you can read the micrometer correctly. As with the point load tester, the main advantages of the cone indenter are that elaborate specimen preparation is not required and that it can be used anywhere in the field.

6.2 Field soil strength tests

The *pocket penetrometer* (Fig. 6.6) is a lightweight and easily carried instrument for measuring the unconfined compressive strength of fine-grained cohesive soils. In use, the penetrometer is pushed slowly into the soil until the penetration mark on the plunger reaches the level of the soil surface. This compresses a spring within the apparatus and the maximum amount of compression is indicated by a pointer which moves along the scale. The scale is calibrated to read unconfined compressive strength directly which is useful, but the units are kgf/cm^2 so that the numbers have to be divided by 10 (strictly speaking 9.81) to give the result in MN/m^2. Because of the small size of the plunger the pocket penetrometer cannot be used on soils containing stones. The surface to be tested should be carefully prepared so that the soil is undisturbed; do not test soil surfaces that have been cut or smeared with mechanical excavator blades. The pocket penetrometer will cover soils within the unconfined compressive strength range 0.05 to 0.45 MN/m^2, that is from soft to very

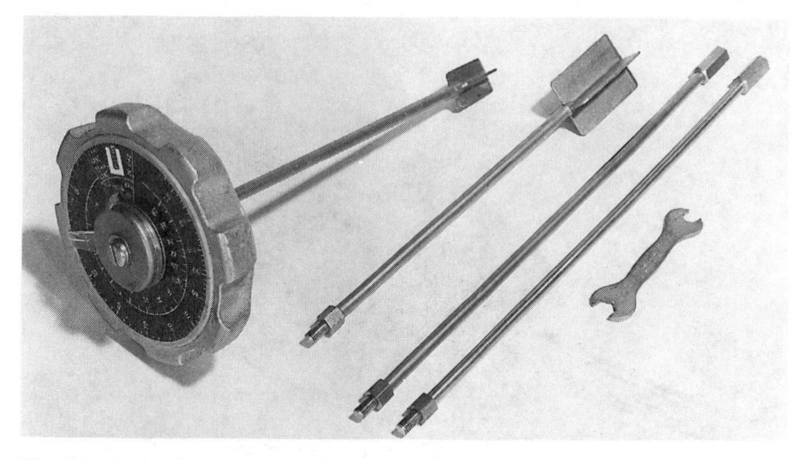

Fig. 6.7 Pocket shear vane, interchangeable vane, extension rods, spanner.

stiff, and can be used for testing *in situ* soils or undisturbed cores either after extrusion or in the ends of the sample tube.

The *pocket shear vane* is a small easily carried instrument for measuring the shear strength of fine-grained cohesive soils. Figure 6.7 shows one version of the instrument. In use, the vane of the instrument is inserted into the soil to be tested and the torsion head is slowly rotated by hand until failure occurs. The maximum shear strength of the soil is read directly from the position of an index pointer carried round a scale engraved on the dial of the torsion head of the instrument. The value should be multiplied by two to give unconfined compressive strength. Different sized vanes are provided for different ranges of soil strength. One of the makes of instrument has extension rods so that tests can be done down a borehole or, for example, through the diaphragm of a tunnelling shield. The vane tester should not be used on stony soils. The shear vane tester will cover soils ranging up to an unconfined compressive strength of 0.25 MN/m^2, that is up to stiff, but is much more suitable than the pocket penetrometer for testing soils of low strength. It can be used for testing *in situ* soils or undisturbed cores in the ends of the sample tube.

The *portable unconfined compression tester* is probably the best field soil strength test but it is less convenient to carry out than the two tests previously described. Like the others, it is suitable for testing fine-grained cohesive soils. The apparatus (Fig. 6.8) consists of a frame in which a compressive force is applied to a cylindrical soil specimen by means of a calibrated spring. The specimen is loaded by turning a cranked handle at the top of the tester and the resulting load–displacement curve for the

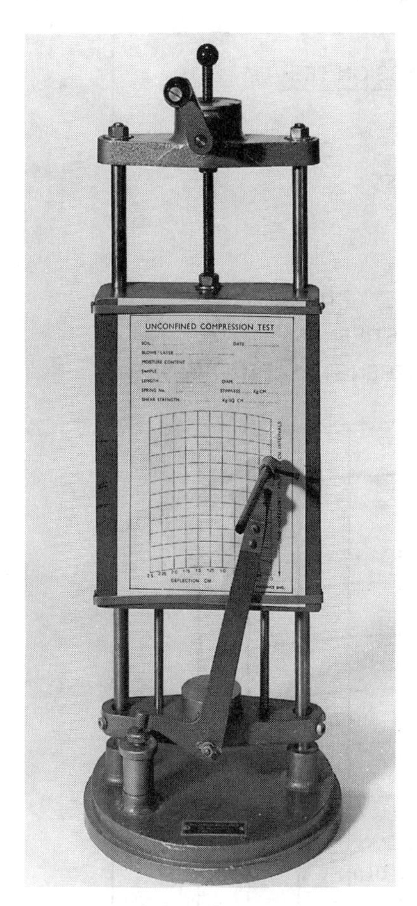

Fig. 6.8 Unconfined compression tester.

specimen is recorded autographically by the movement of a pencil on a chart attached to the front of the apparatus. Special charts are provided for use with the tester (Fig. 6.9), together with a special transparent overlay which allows a parameter δ to be read off at a value corresponding to the peak of the curve on the chart. Multiplying δ by the stiffness factor of the particular spring in use gives the unconfined compressive strength of the specimen. Four calibrated springs are supplied which have stiffnesses of 2, 4, 8 and 16 N/mm extension, and the appropriate spring is chosen to suit the anticipated strength of the soil to be tested. The apparatus is suitable for testing soils ranging in strength up to 1.6 MN/m^2. If the specimen fails along a pre-existing fissure this fact should be noted on the record, such a result being the fissure strength and not the material strength. Specimens 38 mm in diameter and 80 mm long are tested in the apparatus, and to obtain them you require 38-mm diameter undisturbed sampling tubes, an extruder, and a former and knife for trimming them to size. The undisturbed samples may be taken from boreholes, or from the faces of trial pits, trenches, tunnels, etc. A device is also available which enables three 38-mm diameter undisturbed samples to be taken from a U100 open-tube sampler (Chapter 8). The unconfined compression tester, springs, charts, etc. are all supplied in a wooden case, easily transportable by car. Full instructions for using the tester are given in BS 1377:1975, and by the Road Research Laboratory (1952).

Finally, remember that the Mackintosh probe can be used to roughly estimate the *in situ* strength of clays as described in Section 2.1.

In the United States the unconfined compressive strength of cohesive soils in the field is measured using the *Rimac tester*, a device originally developed for testing motor-car valve

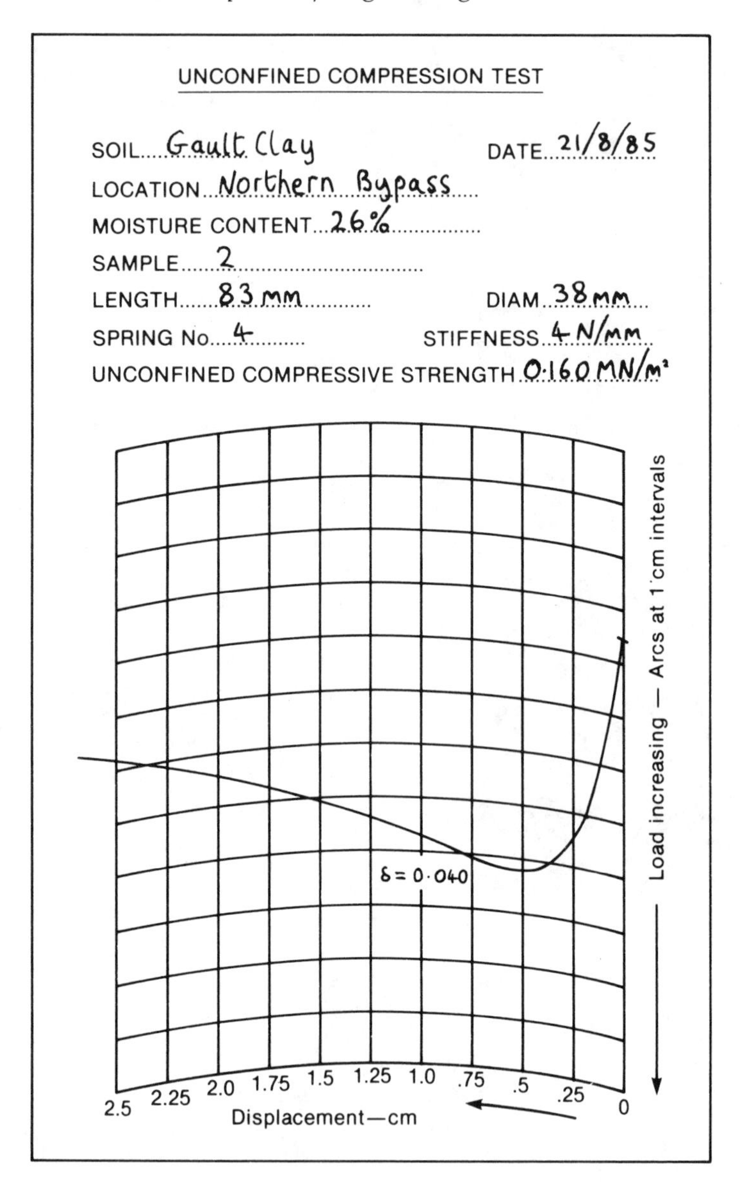

Fig. 6.9 Record of unconfined compressive strength test.

springs. A 35-mm diameter cylindrical soil sample is taken using the split-barrel sampler from the standard penetration test. This is then removed from the sampler, carefully trimmed to length and placed between the platens of the Rimac tester. The original length of the specimen is noted directly from a scale on the side of the tester. A compressional load is then applied to failure and the maximum value of the load is read from a dial graduated both in lbf and kgf. At the same time the specimen's length at maximum load is also noted from the scale. The unconfined compressive strength of the soil is given by

$$\frac{PL_1}{A_0L_0}$$

where P is the maximum load applied, A_0 is the original cross-sectional area of the specimen, L_0 is the original length of the specimen and L_1 is the length of the specimen at maximum load. Care should be taken only to test samples that seem relatively undisturbed when the split-barrel sampler is opened.

6.3 Scale of strength for rock and soil

Table 6.1 shows a universal scale of strength for rock and soil and appropriate terms that can be used for different ranges of the scale. It has been shown how the results of all the tests described in Sections 6.1 and 6.2 can be expressed in terms of unconfined compressive strength so that as well as quoting the test result, the correct term to describe the strength can be chosen. It should be noted that where soils and rocks overlap, the material can be described as a hard soil or a weak rock as seems most appropriate in a particular case. Also shown in Table 6.1 are the ways of estimating strength if the quantitative field testing methods are not available. The scale given in Table 6.1 is the same as that of BS 5930:1981.

Do not expect the different tests, particularly those for rock, to give exactly the same value of unconfined compressive strength for the same material. This is because of the empirical nature of the correlation of the test result and the unconfined compressive strength with all the tests except the unconfined compression testers. The tests are not a substitute for laboratory tests: they are a means of quantifying field observations and getting an early indication of the strength of the materials on site. Remember that some samples will be later tested fully in the laboratory; get the results and compared them with your field test results.

Table 6.1 Scale of strength for rock and soil

Term	*Unconfined compressive strength (MN/m²)*	*Field estimation*
Rock		
Extremely strong	> 200	Very hard rock – more than one blow of geological hammer required to break specimen.
Very strong	100–200	
Strong	50–100	Hard rock – hand held specimen can be broken with single blow of geological hammer.
Moderately strong	12.5–50	Soft rock – 5 mm indentations with sharp end of pick.
Moderately weak	5.0–12.5	Too hard to cut by hand into a triaxial specimen.
Weak	1.25–5.0	Very soft rock – material crumbles under firm blows with the sharp end of a geological pick.
Very weak rock or hard soil	0.60–1.25	Brittle or tough, may be broken in the hand with difficulty.
*Soil**		
Very stiff	0.30–0.60	Soil can be indented by the fingernail.
Stiff	0.15–0.30	Soil cannot be moulded in fingers.
Firm	0.08–0.15	Soil can be moulded only by strong pressure of fingers.
Soft	0.04–0.08	Soil easily moulded with fingers.
Very soft	< 0.04	Soil exudes between fingers when squeezed in the hand.

* The unconfined compressive strengths for soils given are double the shear strengths.
(After Geological Society Engineering Group Working Party, 1977.)

7 Natural and artificial exposures

Existing natural and artificial exposures of rock and soil allow the engineering geologist to examine and sample the geological materials of an area or site without the need for specially made excavations, boreholes, etc. and will be most useful during the reconnaissance stage of a project. However, there are limitations on the use of existing exposures in this way which have to be borne in mind when making such examinations. The main one is that the rocks and soils revealed in existing exposures will have suffered weathering to a greater or lesser extent depending on the age of the exposure so that their properties may be considerably different to the unweathered material. Indeed, the inexperienced engineering geologist should take any opportunity he gets to examine rocks at surface exposure and underground, say in a tunnel, to see how the same rock can differ in appearance and properties when fresh and when weathered.

When examining either natural or artificial exposures, the precepts on field behaviour and safety given in Chapter 1 and Appendix 1 of *Basic Geological Mapping* of this series should be followed. When examining exposures in any working quarries or current engineering works the advice on safety given in Chapter 9 of this book should be followed and any special requirements of the authorities in charge should be complied with.

7.1 Natural exposures

Natural exposures include rock outcrops, cliffs both inland and coastal, stream sections and rocky ridges. If required, exposures of this kind can be mapped as described by the Geological Society Engineering Group Working Party (1972) and in Chapter 4 of *Basic Geological Mapping*. Description of materials, measurement of discontinuities, *in situ* strength testing and sampling can all be carried out using the methods described in Chapters 4, 5 and 6 of this book, but the results will be for material weathered to some degree and may well be different to the properties of the unweathered rocks and soils.

Joints will probably be more open than in the unweathered rock because

Fig. 7.1 Exposure of rhyolite in coastal cliff.

of the action of rainwater and relief of stress. Any calcite filling originally present in joints may have long since been dissolved away. Nevertheless, natural exposures will given some indication of the materials present and are more useful in cold or temperate climates where weathering is less severe than in warm or hot climates. In hot wet climates there may well be *no* natural exposures of unweathered rocks. Natural exposures, of course, suffer from the limitation that they have to be taken where they are found, which may not be close to the engineering project that is planned. Sometimes the reverse is true, and in particular for coastal works, cliffs can provide very fine continuous sections. Figure 7.1 shows measurement of joint spacing underway along a vertical scanline on a coastal cliff exposure, and Fig. 7.2 shows the engineering geologist's sketch of the

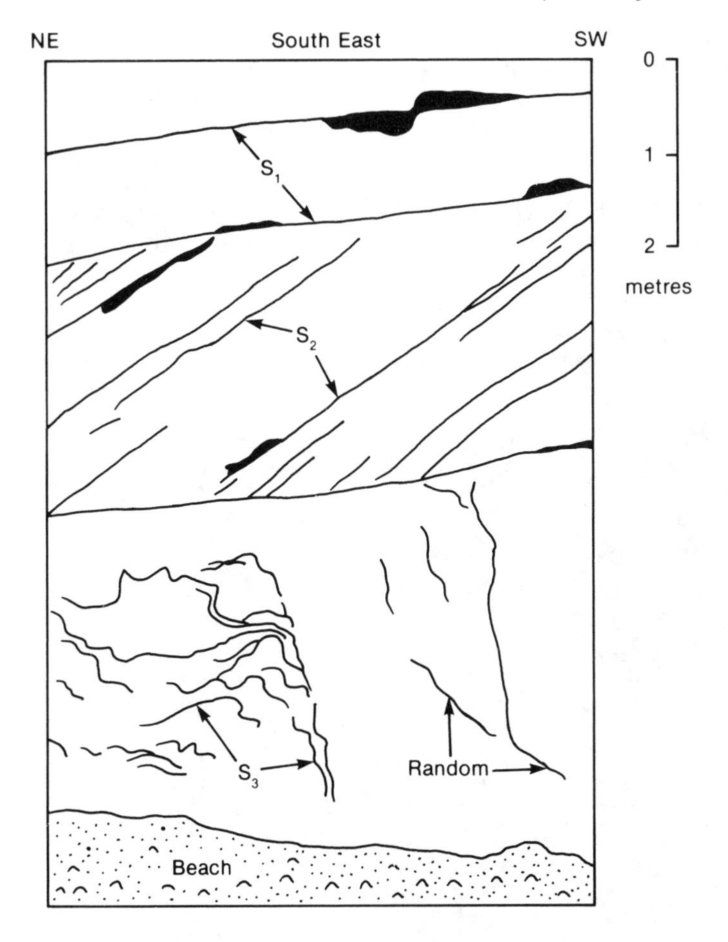

Fig. 7.2 Sketch highlighting different joint sets seen in Fig. 7.1 by selectively omitting all but one set in specific areas. Joint sets are as follows. S_1: Widely spaced, laterally extensive joints, dip 10° to NE. S_2: Medium spaced, laterally extensive joints, dip 30° to NE. S_3: More pervasive near-vertical joints that govern cliff angle. Majority of joints closed or tight, many welded.

same area with remarks on the joint sets. In Fig. 7.1 the observer is safer than he looks in the photograph, the cliff sloping back at an angle.

7.2 Artificial exposures

Artificial exposures include quarries, sand and gravel pits, railway and road cuttings and any temporary exposures for current engineering works or for the installation or repair of services. For some of these, description of materials, measurment of discontinuities, *in situ* strength testing and sampling can be carried out using the methods of Chapters 4, 5 and 6. In addition, for older railway and road cuttings some idea can be gained as to the stability of the soils and the resistance of the rocks to natural weathering. Like natural exposures, artificial exposures have to be taken

Fig. 7.3 Exposure of limestone in quarry face.

where they are found, which may or may not be near the proposed job. An 'entrenching tool', such as can be obtained from campers' stores, is useful for excavating and sampling natural and artificial exposures in soils, but be sure to do this with care and reinstate the site afterwards. Permission should be obtained before any sampling on private or public land. The spoil from rabbit burrows and the like can given an indication of soil conditions where exposures are few (see also Section 7.2.2).

Figure 7.3 shows the typical appearance of a quarry face in massive horizontally bedded limestone, and variations in the bedding and jointing can be well seen; in particular it can be observed that towards the top of the face the homogeneous limestone gives way to limestone interbedded with a softer rock, probably mudstone. Water seepage from the upper layers has produced wet marks on the rock which show up in the photograph as dark tones. An adit, whose portal can be seen at the bottom of the quarry face covered with a shelter, leads to three underground headings as shown in Fig. 9.9. With tall, nearly vertical quarry faces such as this, direct access for sampling and testing the rock is possible only at the bottom unless staging is erected.

7.2.1 *Photographing exposures*

Photographs should be taken of all natural and artificial exposures that are examined or from which soil or rock samples are taken. With very large exposures such as large coastal cliffs or large quarry faces it may be difficult to record the whole face satisfactorily on a single photograph. For example, with a 35-mm camera fitted with a standard lens of 50 mm focal length, the angle of view is only 40°, so that the camera has to be positioned a considerable distance from the face (at point A in Fig. 7.4) if the whole of it is to be included. A way of overcoming this problem is to take a series of slightly overlapping shots from a closer distance (at points B_1 to B_5 in Fig. 7.4), and then trim and stick the prints together to make a panorama of the face. Remember to include a scale in each photograph – ranging rods are very suitable for this purpose – and then measurements of any geological features, bed thickness for example, can be made from the prints afterwards. In fact, if the precaution is taken of always having a scale in a photograph, much simple photogrammetric work is possible, as described by Williams (1969), by utilizing the perspective geometry of the photograph to estimate the size of features in the scene.

Another useful technique is the taking of stereopairs. All that is needed to do this is to take one photograph, move the camera carefully to one side (points A and B in Fig. 7.5), and take another shot of the same scene. The distance AB, known as the camera-base, depends on the distance of the camera from the object of interest. For nearby subjects such as a quarry face or natural outcrop, a

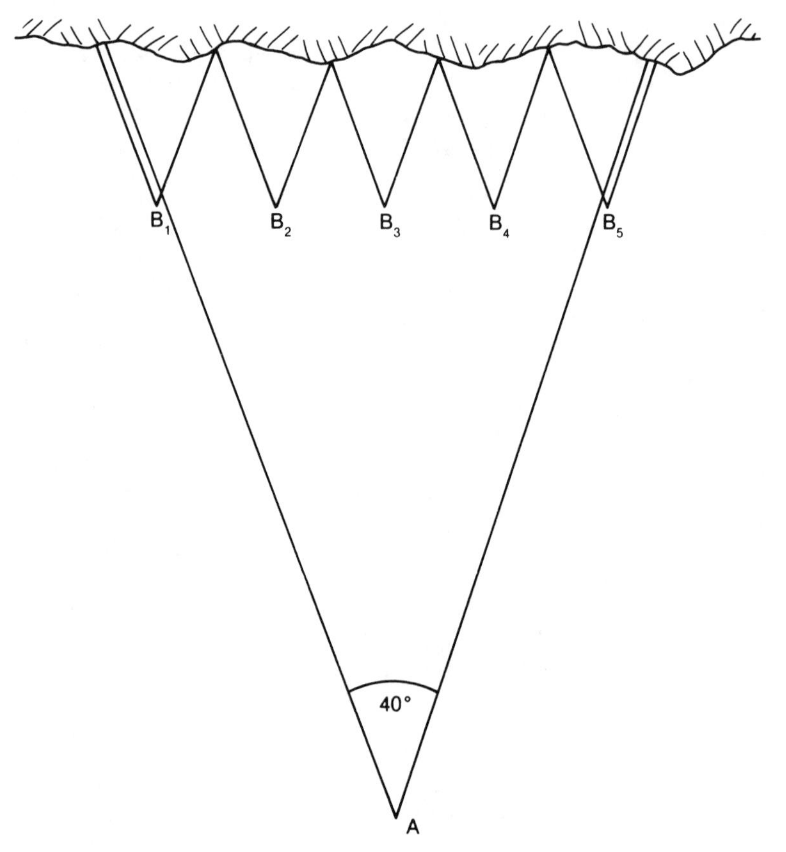

Fig. 7.4 Photographing cliff or quarry face.

spacing between camera stations of 0.5 to 1 m is suitable, but for subjects in the middle distance such as an escarpment slope, a camera-base of several metres or more may be appropriate. Try to avoid including unwanted foreground objects in the photographs because they may be impossible to 'fuse' under the stereoscope and therefore prove distracting. A tripod may be used to assist in accurate framing of the shots, but it is not necessary otherwise. When the prints from these are viewed with a stereoscope, a three-dimensional image of the area of overlap of the photographs will be seen. This technique is extremely useful for photographing slopes because, as with air photographs, the exaggerated relief

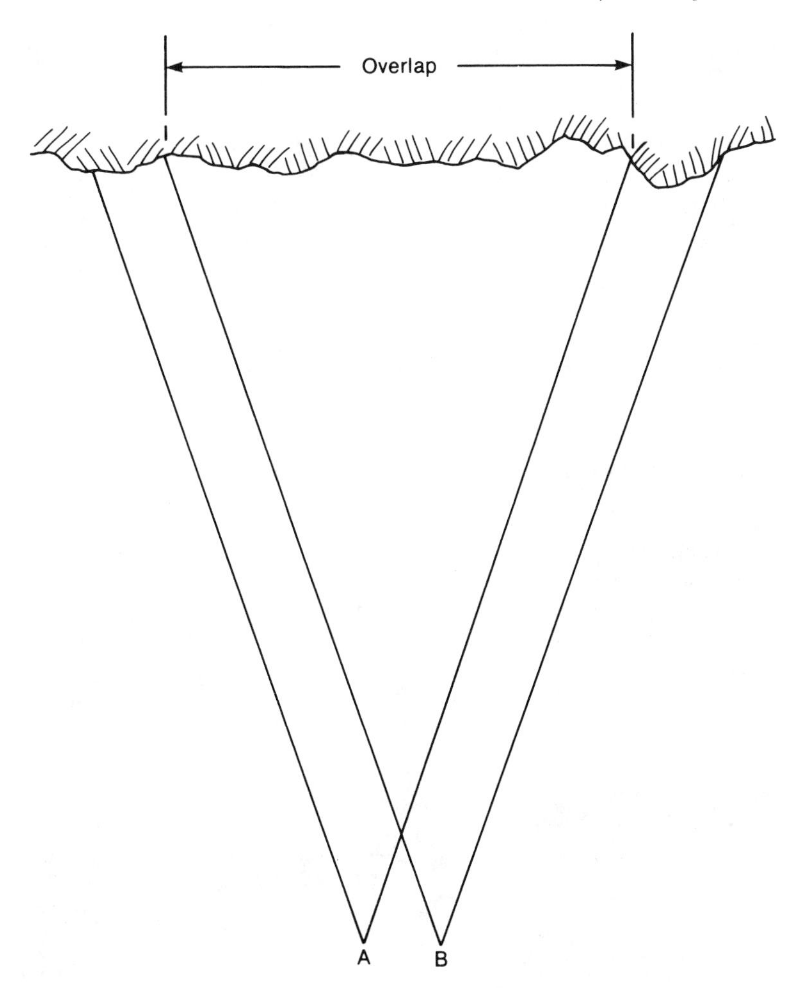

Fig. 7.5 Stereophotography of a face.

produced by the camera-base being much greater than one's eye-base allows very subtle breaks in slope, hummocks, lobes or other signs of slope instability to be detected. Both the techniques of taking panoramas and stereopairs can be applied to some of the situations discussed in Chapter 9, and indeed to the photography of the landscape in general.

For record purposes only, a panorama can be made by rotating the

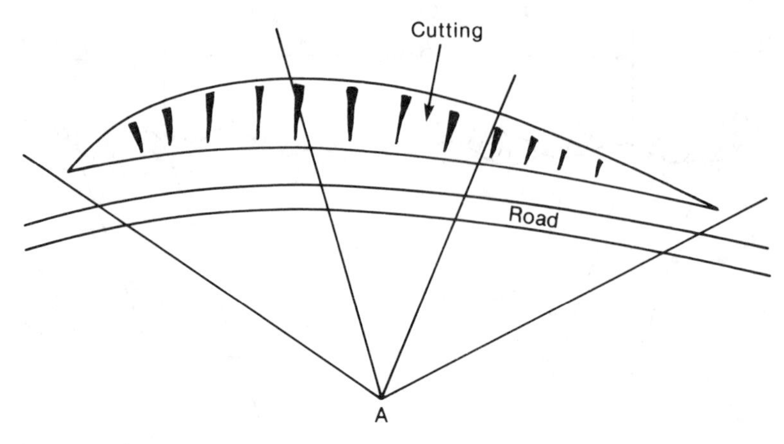

Fig. 7.6 Taking a simple panorama.

camera at a single point, as shown in Fig. 7.6 where the whole of a long road cutting is covered by three photographs from the single position A. As before, the prints are trimmed and stuck together. However, with this technique there is a very large change in scale along the length of the photographs. An example of a two-shot panorama photograph is shown in Fig. 7.7 which records a long length of coastal cliff section. The technique

Fig. 7.7 Two-shot panorama photograph of a coastal section of Chalk cliffs.

Fig. 7.8 Two-shot vertical panorama photograph of Chalk cliff.

can also be used to produce vertical panoramas as shown in Fig. 7.8; the scale is indicated by the rucksack at the base of the cliff but because the camera has to be tilted up to take the upper photograph, the scale will not be the same at the top of the cliff. Panoramas can also be used to make photographic records during construction as illustrated in Figs 9.17 and 9.18.

7.2.2 *Shallow hand-augering*

Between exposures, or in terrain where there are no natural or artificial exposures, a reconnaissance survey can be made by using the technique of shallow hand-augering to get down beneath the surface vegetation and topsoil. For this the auger tool from the boring and prospecting tool kit (see Section 2.1) is employed. The method is to bore down into the soil with the normal rotary action and then to withdraw the auger with a straight pull; the sample of soil retained within the flights of the bit are then removed and examined. Except in gravels, enough soil can be obtained to give a soil material description (Section 4.2), but the mass properties cannot of course be observed. The boring should be done in short increments so that the auger can be pulled out without excessive effort and also to observe any changes in the soil profile. On completion of a boring, reinstate the surface to prevent injury to stock. The technique is extremely useful and rapid, and if you do not have the boring and prospecting kit it is worthwhile having a simple hand-auger made as shown in Fig. 7.9: this is fabricated by letting a length of steel rod into the stem of an ordinary woodworking auger and welding the joints.

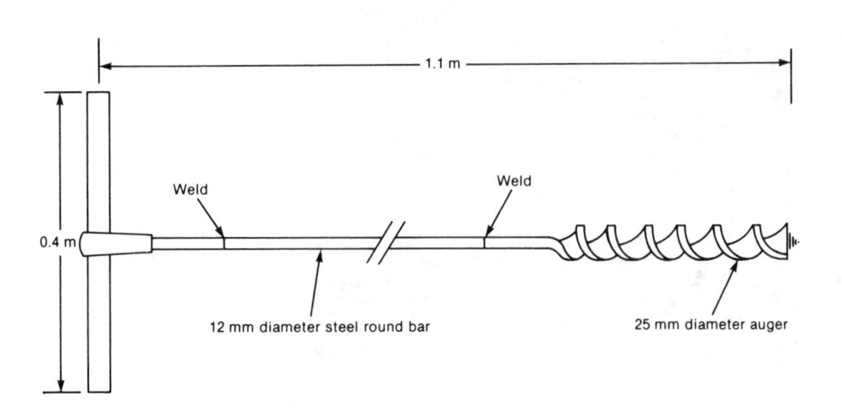

Fig. 7.9 Simple hand-auger.

8
Soft ground boring

Generally, the most suitable method of soft ground boring is the light cable percussion rig (Figs 8.1 and 8.2) using the shell when boring through granular soils and the clay cutter when boring through cohesive soils. It should be noted that in civil engineering, the light cable percussion rig is almost always referred to as the 'shell and auger rig', this name being a hangover from the days when an auger instead of a clay cutter was used for boring in clay. The borehole can be supported with casing if required and boulders and cobbles can be broken up with the chisel. Undisturbed samples of cohesive soils can be taken, and standard penetration tests in granular soils can be carried out. Some light cable percussion rigs can be quickly adapted to rotary coring if the borehole should run out of soft ground into rock. Some of these rigs can be folded up and towed like a trailer behind a car.

The driller in charge of an individual rig is normally responsible for recording all the information obtained from the borehole; this is usually done on a standard daily report form, often referred to as the driller's journal. He is also responsible for taking samples, carrying out standard penetration tests and noting ground-water levels. Any special sampling or testing in boreholes over and above that done as routine is carried out by field technicians. Sometimes, the young engineering geologist will be asked to make a field description of the strata the borehole passes through by describing the material brought up by the shell or clay cutter during boring, to record the taking of disturbed samples, undisturbed samples and standard penetration tests, and to record observations of any ground water encountered during boring. However, the experienced engineering geologist may be called upon to supervise the whole field programme, so take every opportunity of finding out about all aspects of the work. The frequency of sampling will depend on the purpose of the borehole but the following is typical of a borehole for foundation design. As a general rule, undisturbed samples of cohesive soils will be taken every 1 to 1.5 m with disturbed samples being taken between them. In addition a disturbed sample immediately followed by an

Fig. 8.1 The Pilcon Wayfarer light cable percussion rig. (Courtesy of Pilcon Engineering Ltd.)

undisturbed sample is usually taken at each change in soil type or stratum. The 100-mm diameter open-tube sampler, commonly referred to as the U100 or the U4 (its diameter in inches), is used for taking undisturbed samples in firm to stiff clays, but a thin-walled sampler will be used for soft clays and silts. In granular soils, large bulk samples are taken at 1 to 1.5 m intervals with small jar samples in between. Again, in addition a sample is taken at each soil or stratum change.

The relative density of sands and gravels is determined by the standard penetration test (SPT) in which the number of blows to drive a standard tool a depth of 300 mm is recorded; this is known as the *N*-value. The interpretation of *N*-values in terms of relative density is given in Table 8.1.

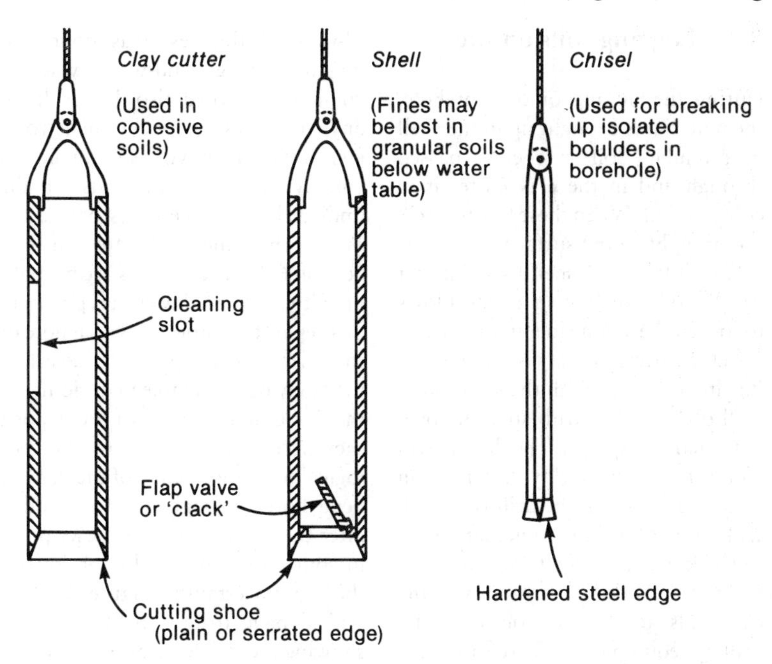

Fig. 8.2 Tools used in cable percussion boring.

Table 8.1 Relative density scale for sands and gravels

Term	*SPT* N*-values: blows/ 300 mm penetration*
Very loose	0 to 4
Loose	4 to 10
Medium dense	10 to 30
Dense	30 to 50
Very dense	over 50

(After BS 5930:1981)

8.1 Logging soils on site

During the sinking of the borehole, the material is brought up in the shell if granular soil is being passed through and in the clay cutter if in cohesive soil. When these boring tools are brought to the surface, the soil is removed from the shell by turning it upside down and giving it a few blows to loosen the soil inside; in the case of the clay cutter, the clay is removed by digging it out through the slots in the wall of the cutter with specially provided narrow spades. The drillers will empty the shell or clay cutter on an area to one side of the drill rig which will soon become an unpleasant place to work, particularly if wet. Also, the rest of the immediate area around the rig tends to become occupied by drilling equipment. Therefore, provide yourself with a set of metal trays (the 300 mm stainless-steel trays (see Fig. 4.1) used in soils laboratories are ideal) which you can use to collect your samples and these can then be taken to a clean uncluttered area where you can work on them. Make sure that the samples are taken from the lower part of the shell or clay cutter so that there is no danger of them being contaminated by material which may have fallen in from higher up the borehole.

The samples should be described using the method outlined in Section 4.2 and the descriptions entered directly into your field notebook or onto a standard form. Record the depth, the nature of any samples taken and their numbers, the SPT test *N*-value if this test was done and whether any ground-water was encountered. Record also the results of any shear vane or pocket penetrometer tests which you can make on cohesive soils in the ends of the undisturbed sample tubes before the drillers put the end caps on (see Section 6.2). If any fossils are brought up in the shell or clay cutter, put them in a separate sample jar with a note of their depth; when properly identified they may be invaluable in deciding on the statigraphical level of the deposit they occur in. If you know the geological formation name of the deposit being bored through this can be added after the soil name in the description in your notebook but do not do so if there is uncertainty. Figure 8.3 is a typical page from a field notebook showing the soil description and other information recorded by an engineering geologist at a borehole during drilling.

When working around a boring rig, always wear a hard hat and safety boots, and if you have to handle the shell and clay cutter wear a pair of strong gloves. Remember too that the shell and clay cutter are very heavy, particularly when full of soil.

8.2 Logging soils off site

On other jobs, however, the driller in charge of the rig will do all the site logging himself except that later on, the engineering geologist will be called upon to provide the soil description by examining the disturbed

Borehole 12 cont. 23/6/82

11·50 – 11·90 m

Very dense, unweathered, greyish green, glauconitic, slightly clayey, fine to medium SAND. A few shell fragments present. (Bracklesham Beds).

Jar sample No. 375

11·95 – 12·40 m

SPT test with split-barrel sampler N = 62

(4, 9, 11, 14, 17, 20)

Fig. 8.3 Page from a field notebook. Note the geologist has recorded the kind of tool used for the SPT test (split-barrel sampler or solid 60° cone for use in sands and gravels, respectively) and the number of blows to drive the tool each increment of 75 mm including the first two which are the seating drive and are not included in the *N*-value.

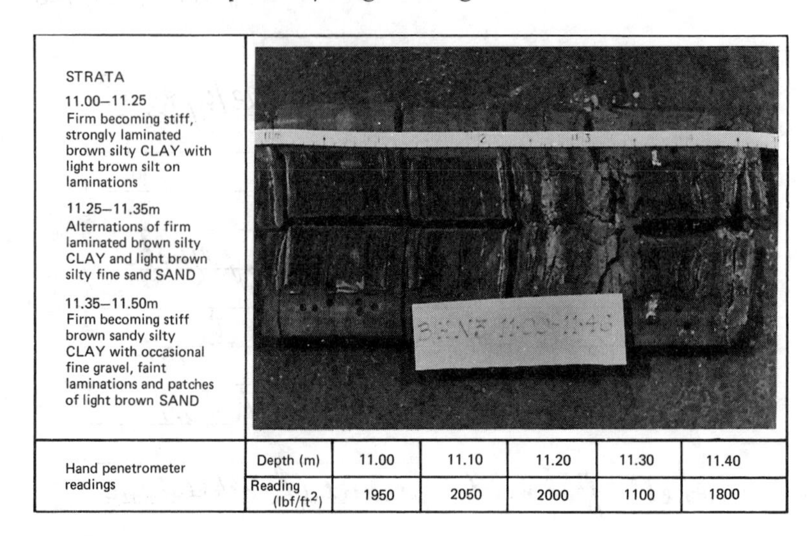

STRATA

11.00–11.25
Firm becoming stiff, strongly laminated brown silty CLAY with light brown silt on laminations

11.25–11.35m
Alternations of firm laminated brown silty CLAY and light brown silty fine sand SAND

11.35–11.50m
Firm becoming stiff brown sandy silty CLAY with occasional fine gravel, faint laminations and patches of light brown SAND

Hand penetrometer readings	Depth (m)	11.00	11.10	11.20	11.30	11.40
	Reading (lbf/ft^2)	1950	2050	2000	1100	1800

Fig. 8.4 Split core photograph and record sheet giving soil descriptions and pocket penetrometer readings.

samples in the sample store or laboratory; this will, in general, be less satisfactory than being at the borehole at the time. You will need a well-lit bench to work on, some metal trays, a glass plate, palette knives and a supply of water. Disturbed samples are usually stored in large screw-topped glass jars. Empty the whole of the sample into the tray and describe the soil material using the methods of Section 4.2. Record the borehole number, sample number and depth along with the sample description. Do not forget to replace any inner label when returning the sample to the jar. The undisturbed samples will be extruded in the soils laboratory where specimens for strength and consolidation tests will be prepared from them. However, any unused cores, or unused lengths of cores should be split open longitudinally by cutting along one side with a palette knife and then pulling apart, for a detailed examination of the structure. Fissures, bedding plane partings, joints and any major slip planes present should be looked for, noted and a sketch made of them. They should be described using the methods of Section 4.1. Penetrometer or shear vane tests can be made (see Section 6.2). The split-opened core should be photographed (Fig. 8.4), preferably in colour; remember to include a scale in the photograph and a standard photographic colour-card as well if one is available. It is a very simple matter to take stereopairs of the surface of a split-opened soil core: Figure 8.5 shows the arrangements and all that it

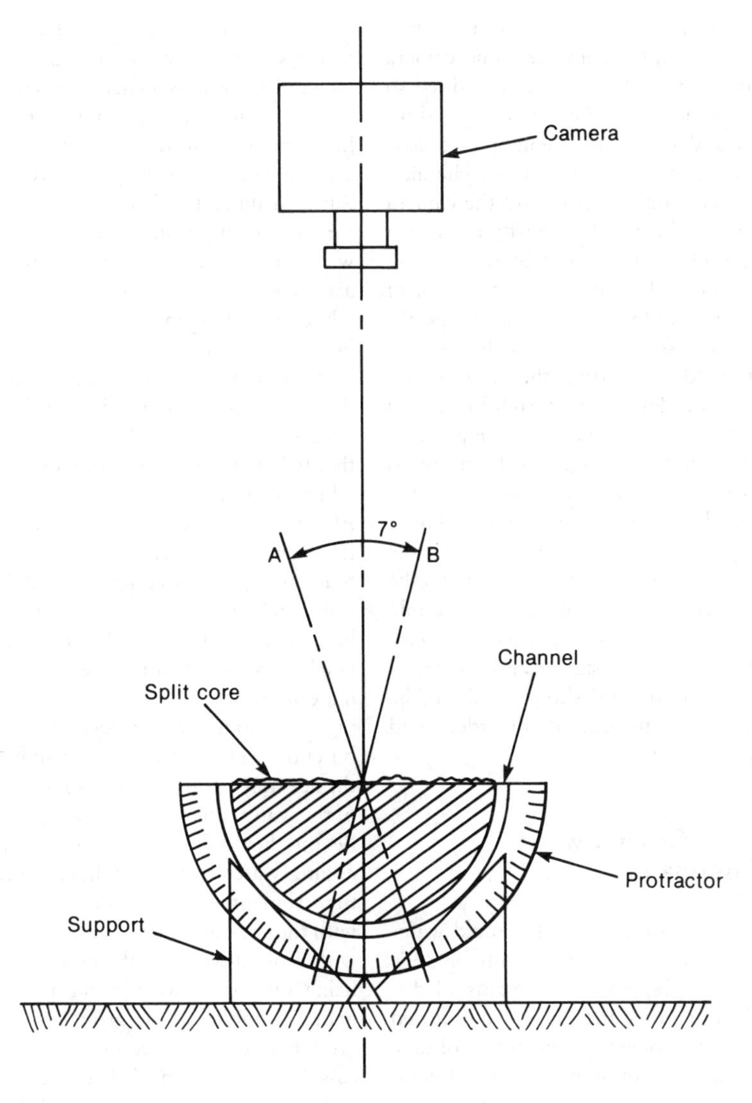

Fig. 8.5 Arrangements for taking stereopairs of a split-opened soil core. A tilt of 7° gives a good stereoscopic effect.

is necessary to do is to take two photographs from the same camera position with the split core tilted so that the perpendicular is at positions A and B in turn. Lengths of plastic guttering make convenient channels for holding the core and the camera should be fixed in position using a tripod or retort stand. Stereopairs are useful if the core shows a wealth of structural features. If you are specifically looking for a major slip surface, instead of splitting the core as previously described, it should be laid intact on the bench and the plane of discontinuity sought for by trying to shear the core sideways with the hands using gentle pressure. For a failure or instability investigation the use of core in this way may take precedence over its use to supply laboratory test specimens. Once found, the position and angle of inclination of the slip plane should be carefully measured, recorded and photographed.

8.3 Ground-water information

The determination of ground-water conditions at a site can only be done properly if piezometers are installed in the boreholes and they are monitored over a sufficiently long period of time to take account of seasonal fluctuation, and this kind of investigation is beyond the scope of this book. Valuable ground-water information can, however, be obtained from the borehole during drilling, and the driller in charge of an individual rig has the responsibility of recording ground-water observations on his daily report form. The information that the driller has to record is shown in Table 8.2. If the engineering geologist is present during drilling it will not be necessary for him to duplicate all the driller's water observations. However, there are three observations that are worthwhile recording in your field notebook because of their importance in understanding ground-water conditions. Firstly, record the depth at which water is struck during boring – this tells you the stratum or deposit which is the source of ground-water. Also note if water is struck at more than one level, there may be more than one aquifer. Secondly, record the rate at which water rises in the borehole after it is first struck – this tells you how permeable the water-bearing material is. Thirdly, record the overnight standing water level in the borehole before drilling is resumed the next day – this gives you some idea of the water table level for the season. Be on the look-out for the drillers adding water to the borehole which they sometimes do to assist boring with the shell and do not confuse this with ground-water. If the borehole is being sunk for a foundation investigation, samples of the ground-water are taken to determine if any chemicals deleterious to buried concrete or metal are present; it is essential that these samples are not contaminated with water added during boring.

Recording ground-water information during rotary drilling in rock

Table 8.2 Water information recorded by driller on daily report forms

Water record on daily report form of cable percussion boring					
Water levels, m				on pulling casing	24 h after pulling casing
Time of day					
Depth to water					
Depth cased					
Depth of hole					
At what levels was water encountered?					
Did the level rise?					
If so, how much and how fast?					
Was water added to assist boring?					
If so, at what depths?					
At what depths was water cut off by casing?					
If standpipe or piezometer inserted, to what depth?					
NOTE. If more than one water level is encountered give details of them all.					
Additional items on water record on daily report form of rotary drilling					
Colour of water return.					
At what depths was full circulation not maintained and state percentage return at these depths?					
NOTE. If more than one water level or circulation loss is encountered give details of them all.					

(Adapted from BS 5930:1981)

is more difficult because water or drilling mud is usually used to flush the drilling debris out of the borehole and the presence of this water or mud precludes observations on the ground water. However, if water flush is used, a watch should be kept on the colour of the water return – a change in colour may indicate inflow of ground water into the borehole. Any gain or loss of circulation should be noted, the former indicating inflow of ground water into the borehole and the latter that voids in the rock may have been encountered.

8.3.1 Measurement of ground-water level

To make the measurements of ground-water level referred to above, a water level meter (sometimes called a dip meter) is used. This is an instrument that consists of a graduated electrical cable with a weighted probe at the end, wound on a handreel containing an audible or light indicator and battery (Fig. 8.6). Different cable lengths are available from 30 to 500 m, some are graduated every metre and some every centimetre. A very compact pocket dip meter is also available, but it has only 10 m of cable. To use the instrument, the probe is lowered down the borehole by slowly unreeling the cable: when it touches the water surface an audible or a light signal is given. The length of cable is then read off at the top of the borehole to give the depth to the water surface. The driller in charge of the rig will usually have a water level meter

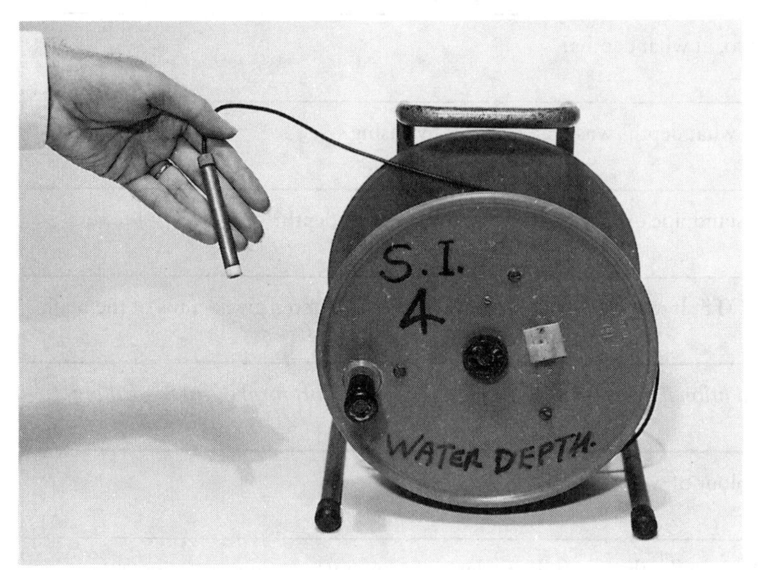

Fig. 8.6 Water level meter.

but you should have your own so that you can measure the water levels in completed boreholes as well as during boring. Before using the instrument, check that it is working satisfactorily by dipping the sensor of the probe into a bucket of water and seeing that the indicator signals correctly. Also, and most importantly, make periodic checks using a surveyors' measuring tape to see that the graduation markers on the cable are in their correct positions, and if not re-position them.

8.3.2 pH measurement of ground-water

Samples of ground water taken during the course of boring will be tested, if required, back at the main laboratory. The only test of ground water that it is usual to carry out in the field is the measurement of pH, which can be done using a portable colorimetric soil testing outfit. A sample of the ground water, which should be tested as soon as possible after it is taken and must not be contaminated with drilling water, is collected in a clean glass or plastic bottle of about 1-litre capacity. If necessary, allow the bottle and contents to stand until any suspended sediment has settled out. A quantity of the clear water is then poured into one of the glass tubes supplied with the testing outfit until the water level reaches the first graduation mark. The level is brought up to the second graduation mark by adding the special indicator solution and the tube is closed and shaken. The colour of the liquid in the tube is then compared with a colour chart provided with the outfit, enabling the pH value of the ground water to be read off to the nearest 0.5 pH unit.

As its name indicates, the soil testing outfit is also used for testing the pH of soil samples, and directions for doing this are provided with the outfit; a description of the procedure is also given in BS 1377:1975.

9
Excavations

This chapter deals with the engineering geology of civil engineering construction sites which can sometimes be dangerous places so that at the outset some remarks will be made about general safety; the hazards specific to particular kinds of site will be discussed at the appropriate place. The young or inexperienced engineering geologist will need to be on guard against three things: (i) unawareness of dangers because of inexperience, (ii) the belief that 'it won't happen to me', (iii) the attitude that it is somewhat feeble to be concerned about personal safety. The first can be remedied to some extent by reading the British Standards that will be referred to here, and the second by common sense. The third is sometimes prevalent among the workforce on civil engineering sites but the young engineering geologist should not be influenced by it, and should be guided by the advice given here and a proper concern for personal safety. Note that under the Health and Safety at Work Act, 1974, although, of course, employers have their obligations, employees have a responsibility for ensuring their own safety. In the United States, follow the instructions of Construction Industry Standard OSHA 2207.

Always wear a safety helmet, invariably referred to as a 'hard hat', either the civil engineering and mining type or, particularly if working underground, the climbers' type, which is of heavier construction and has a strong permanently attached chinstrap, may be preferred. Whichever type of helmet is chosen, make sure it is fitted with a clip at the front and a snap fastener at the back for attaching a cap lamp and cable for use underground. The civil engineering and mining type of safety helmet must be replaced every three years. Always wear safety boots, that is boots having steel toe caps. They are available both as wellingtons and as leather boots. Leather boots should have rubber soles with a good deep tread pattern (commando-type soles).

Provide yourself with a small rucksack which you can wear on your back and in which you can carry all your items of field equipment, thereby leaving your hands free – this is essential if ladders have to be climbed, as they often do on site. Overalls are convenient to wear for site work but you will need a donkey jacket as well in cold weather and oilskins if working in the rain. A strong leather belt fitted with straps for carrying a cap lamp battery will also be needed – the belt will also be used for carrying your hammer in a hammer frog.

In the older industrial countries, civil engineering construction is becoming increasingly sited on reclaimed land which may include areas where hazardous or toxic industrial wastes have been tipped or disposed of in the past. If called upon to enter an excavation in such a site the engineering geologist should be on the lookout for any dangerous materials, which may be solids such as asbestos waste, liquids such as the residues from old gas works, or gases such as hydrogen sulphide. If in any doubt do not enter the excavation until specialist advice from a chemist has been obtained. Similar precautions are necessary when handling any suspect material brought up from a borehole, either on site or later in the core store or laboratory. A British Standard *Code of Practice for the Identification and Investigation of Contaminated Land* is in preparation and you should consult this when it is published. On old landfill sites backfilled with domestic refuse, there is the particular danger of methane gas. Examples of some contaminated land sites and materials are given in Table 9.1.

There are basically two kinds of logging that can be carried out irrespective of the type of exposure that is being logged and these are general logging and specific logging. In *general logging* the objective is to make a record of all the engineering geological features that are present in the exposure, to describe all the soil or rock types present and to record all the structures, water conditions, depth of weathering, etc. This is the commonest form of logging undertaken during the site investigations for large civil engineering projects where all engineering geological observations may have a significance that cannot be predicted in advance so that it is vital to record everything while the exposure, say a trial pit, is still accessible.

In *specific logging* only certain predetermined information is recorded and this is always done for a special purpose. An example might be the recording of rock strength measurements at a tunnel face once a day during a tunnel drive because the Engineer and Contractor have agreed that this parameter alone shall be used as the basis of deciding on payment for excavation. However, the engineering geologist should not miss opportunities to record other features of importance such as faulting, jointing, changes in lithology, etc. that can be quickly logged and which may have a crucial bearing on ease of excavation and which, therefore, may need to be considered in the event of a dispute or claim. So remember, even if you are engaged on specific logging, do not shut your eyes to other information which may be relevant to the matter in hand.

Finally, whenever on a civil engineering construction site, the young engineering geologist should take

Table 9.1 Contaminated land sites and materials

- Made ground and reclaimed river foreshore
- Metal mining, processing, and plating works
- Ferrous and non-ferrous metal smelters
- Foundries
- Scrap dealers and car breakers' yards
- Tanneries
- Sewage works
- Docklands
- Gas works
- Chemical works
- Re-processing plant for gas industry waste
- Railway goods depots and coal storage yards
- Landfill sites
- Munitions manufacturing and testing works
- Oil refining and blending works
- Transport depots
- Paint works
- Dumping of demolition rubble
- Disposal of ash from heating plant
- Burning of domestic and other wastes
- Raw material stockpiles
- Industrial premises damaged by wartime bombing
- Unexploded bombs
- Domestic refuse tips
- Unburnt and burnt colliery shale tips
- Other waste from mining activity
- Power station ash disposal sites and deposits

every opportunity of broadening personal experience. To follow a job through from site investigation to construction will teach lessons on geological interpretation that cannot be learnt any other way. Find out how different pieces of civil engineering plant, tunnelling machines, etc. perform in different soil and rock conditions. Observe what are the engineering geological factors that are important for different kinds of civil engineering project. Note the overall engineering geological properties of important geological formations (e.g. Coal Measures, Oxford Clay) – and how they may vary from one region to another. Cultivating habits of enquiry and observation of this kind will, in the long term, provide you with a background of experience that will make your judgement and advice valued by Engineer, Client or Contractor.

9.1 Trial pits

One of the most common occupations of the engineering geologist is the logging of trial pits (Fig. 9.1). These are typically up to 5 m deep and are often excavated with a backacter. It is essential that trial pits are excavated so that one side is either gently sloping or formed with a series of shallow broad benches – this is to provide a means of safe entrance and egress for the persons logging in the trial pit. The remaining sides must be made secure against collapse as required by BS 6031:1981.

Trial pits should be logged immediately after they have been excavated. If you have any doubt about the safety of the trial pit do not enter it but log it from the surface only. Assuming that the pit is safe to enter, the first thing to do is to make a scale drawing of the exposed faces, noting particularly the structure of the deposits including any slip or shear planes. If the trial pit is large, it is helpful before making the drawings to attach a grid of tapes to the face – say at 1 or 2 m intervals – to help in producing an accurate record. Tapes can be attached with long nails or with large staples which you can make yourself from stiff wire. Squared paper is useful for making the drawings of trial pits faces on. The faces should then be photographed, preferably in colour. Remember to place a large piece of card with the site name, the trial pit number and the date written boldly on it (e.g. with a *Valve Marker*) in the field of view because all trial pit photographs look much the same several weeks afterwards when you get the photographs

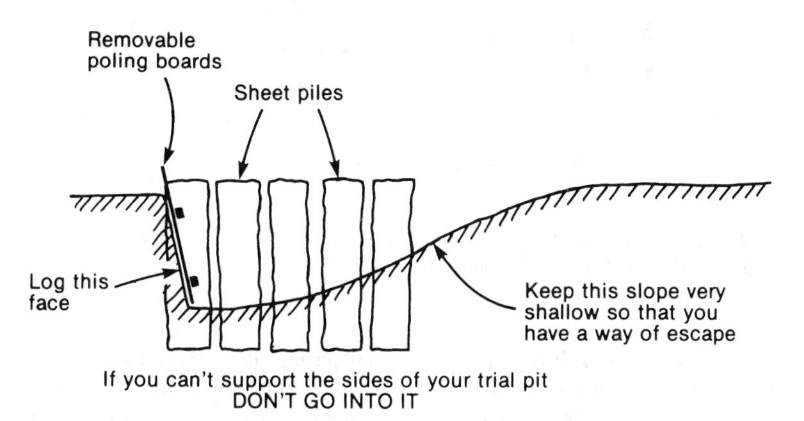

Fig. 9.1 Shallow trial pit (diagrammatic).

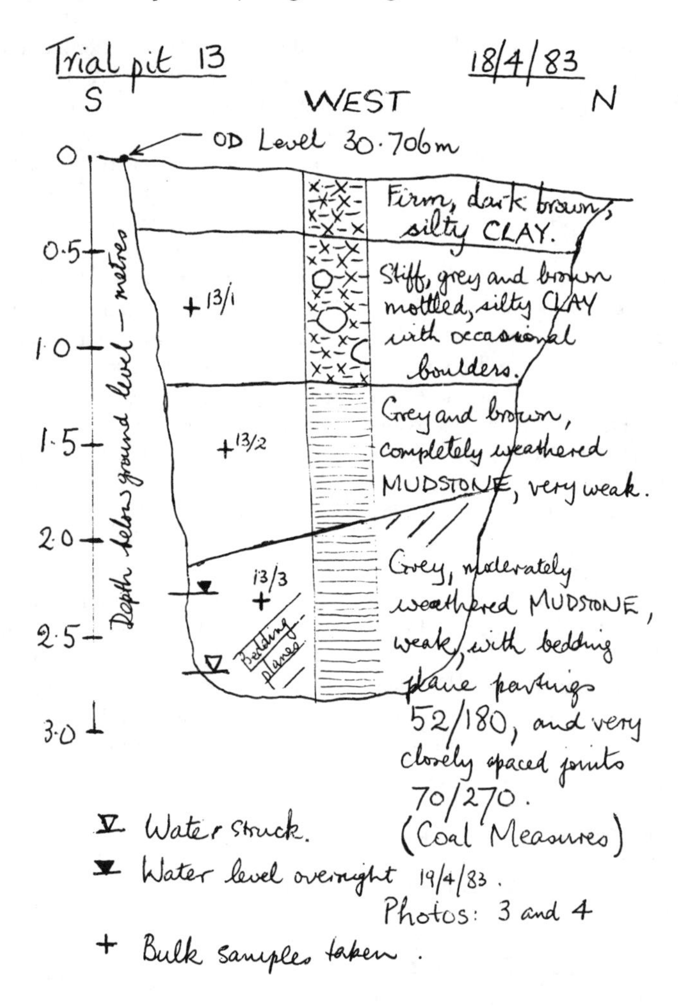

Fig. 9.2 Page from a field notebook showing the log of a face of a trial pit. Note that sampling and ground-water information have been included.

back! Be generous with the numbers of photographs you take because the cost of the film will always be trifling compared with every other expense of trial pitting. Remember to include a scale (a surveyor's ranging rod is convenient) in each photograph. Each stratum or soil type exposed in the trial pit should then be described using the methods of Chapters 4 and 5 (Fig. 9.2), and soil strength measurements should be made as described in Section 6.2. Samples of each material should be taken, the sampling positions recorded on the drawings of the trial pit faces and sample numbers recorded in your field notebook.

Any horizons or levels at which water is observed seeping into the pit should always be noted. If the site is on sloping ground, major slip planes parallel to the ground surface should be specifically looked for. Before making any detailed observations on trial pit faces any disturbed material or smeared surface produced by the backacter bucket should be carefully removed until you are sure you are looking at the original structure. Logging a trial pit from the ground surface will be a less satisfactory procedure than logging the pit directly but is preferable to being buried by the collapse of an unsafe face. Get the backacter operator to use his bucket to dig out samples for you from parts of the faces where you want them and then make your soil descriptions from these. Record in your field notebook that the trial pit has been logged in this way.

9.2 Trenches

The logging of trenches is very similar to the logging of trial pits – a trench can be thought of as a long thin trial

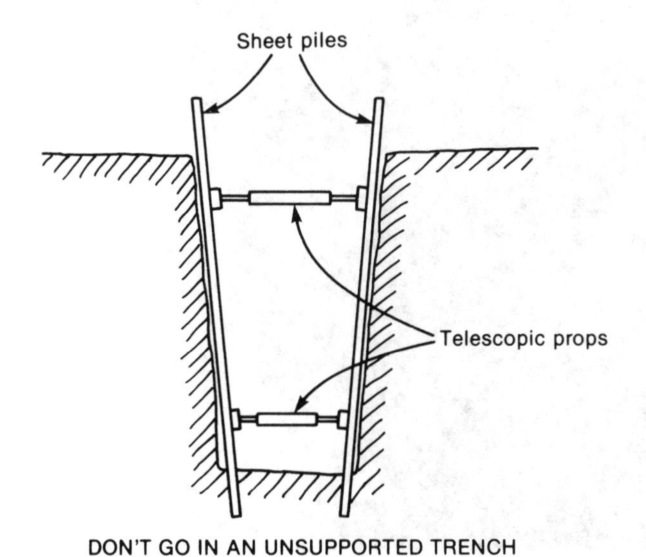

Fig. 9.3 Trench (diagrammatic).

pit (Fig. 9.3). If the trench is over 1.2 m deep it must be properly supported to prevent collapse; BS 6031:1981 and CIRIA Report 97 give details of what is required. Trenches are excavated with a backacter or by hand and the support can be of timber, of steel or a combination of both. The support in deep trenches will generally interfere with the logging of the trench walls and special provision will have to be made when it is installed for certain of the boards or sheets to be removable so that the engineering geologist can log the ground by having them removed and replaced in turn whilst he proceeds along the trench. Shallow unsupported trenches will not, of course, have this difficulty.

The methods used in trench logging are exactly the same as those described for trial pit logging in Section 9.1 and will not be repeated here.

It is useful to attach a surveyor's tape to the end of the trench and run the tape out along the length of the trench to give you a linear scale of

Fig. 9.4 Taking undisturbed samples from the end of a properly supported trench.

position against which to compile your log. Other tapes can be used to make a grid if required, and ranging rods set out at intervals will give a vertical scale. Photography of trench faces may be more difficult than for those of trial pits because of the restricted width and poorer lighting. A short-focus or wide-angle lens may be useful, together with a flash gun, to overcome these difficulties. Trenches and trial pits often provide good positions for the taking of undisturbed samples for strength, consolidation and other soil tests, and the engineering geologist may have to take them or supervise the taking of them. Be sure that any disturbed material is removed before taking these samples. Figure 9.4 shows 38-mm diameter undisturbed samples for unconfined compression testing (see Section 6.2) being taken from the end of a trench. Note the support for the trench which is made from steel sheet piles, timber walings and *Acrow-prop* struts. Note also the ladder for safe access to the trench. Strictly speaking, an operative should not stand on the walings as shown in the photograph, or on any other of the excavation supports. As for trial pits, note the levels and positions at which any ground-water seeping into the trench occurs.

9.3 Man-access boreholes

The logging of man-access boreholes will usually be done as part of the site investigation for the foundations of a large, deep or specialized structure. The logging methods used will be those for rock since any part of the borehole through soil will be cased and therefore not available for examination. It is essential that a properly constructed cage be provided (Fig. 9.5) and that all the detailed provisions of BS 5573:1978 be strictly followed. The following measures must be taken: a banksman must be in attendance at the top of the hole at all times when someone is in the borehole and an alarm, a system of communication, an air supply and lighting must all be provided (also see Table 9.2).

If the engineering geologist who is to log the hole is inexperienced, an experienced supervisor should ensure the hole is safe to enter by descending first. There are special limitations on time, size of hole, precautions against water-bearing and unstable strata, etc. and it is essential that BS 5573:1978 be followed in all these respects.

Having established that the borehole is safe to enter, the geologist should be slowly lowered in the cage to the bottom, making a quick reconnaissance of the borehole as he descends to get a feel for the kind and distribution of features in it. Whilst doing this, a surveyor's tape, attached to the top of the hole, should be unreeled to provide a depth scale down the hole. The geologist should then instruct the crane operator or winchman to raise the cage in increments of say 2 m at a time. At each position the geologist compiles a log of the borehole wall, carries out any *in*

Fig. 9.5 Cage for examining man-access borehole. Note casing, tape, air hose, and power supply for lighting. (Courtesy of T. I. Longworth, Building Research Station.)

Table 9.2 Checklist of safety items before descent of large-diameter boreholes

- Air testing apparatus
- Safety helmet with chin strap and cap lamp
- Safety harness
- Rescue line
- Breathing apparatus and spare air cylinders
- Lighting arrangement
- Communication with surface
- First-aid kit

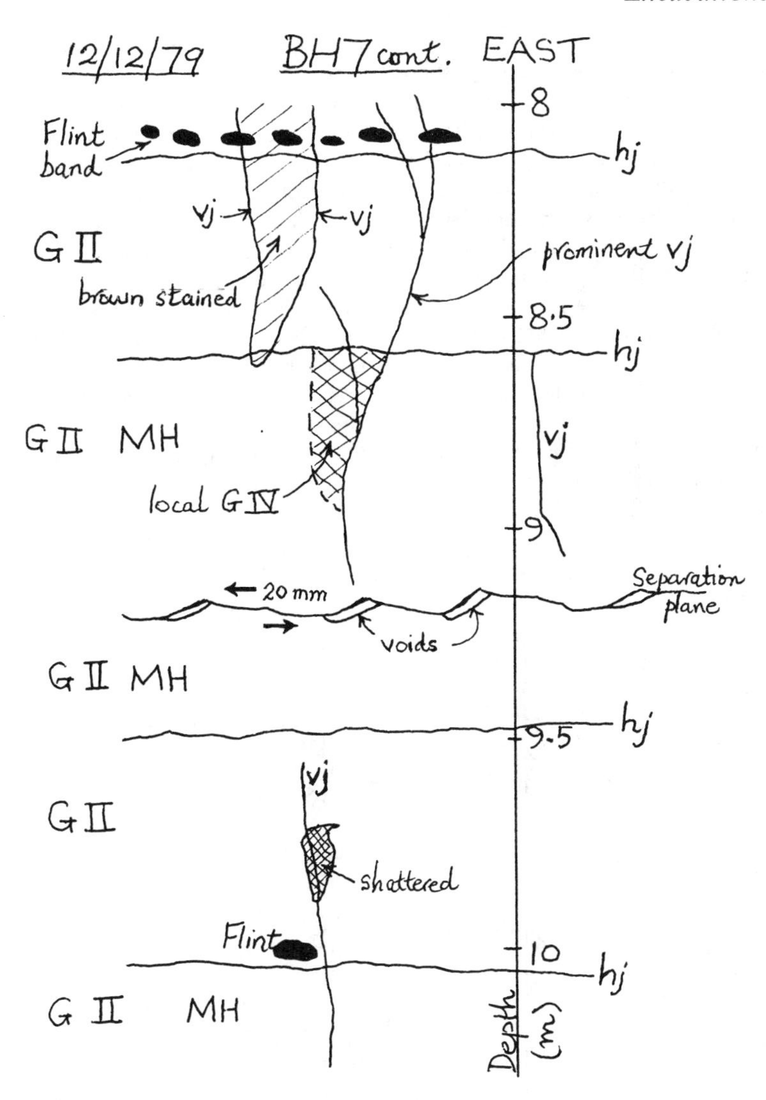

Fig. 9.6 Page from field notebook (simplified) of logging a man-access borehole in chalk. (Courtesy of T. I. Longworth, Building Research Station.)

situ tests and takes any samples required; the methods used are those described in Chapter 5, and Schmidt hammer tests (see Section 6.1) can also be carried out. Don't forget that photographs can be taken but a camera with a short-focus lens is needed if the borehole diameter is small, and a flash gun will be required for illumination. Any water seeping into the borehole should be noted.

There are two ways of compiling the log of the borehole wall. The first is to represent the whole 360° of the borehole wall by the page of the field notebook or log sheet, and this is the most appropriate method if there are numerous geological features revealed, particularly if they are structural features such as bedding planes, faults, etc. The second is to represent only half the borehole wall (180°) by a page, and this will be sufficient in simple geological situations. Whichever method is used, a vertical line should be drawn down the page to indicate a particular compass direction, which should be stated; this is essential for later interpreting the effects of the orientation of the structural features logged.

Figure 9.6 shows a typical page

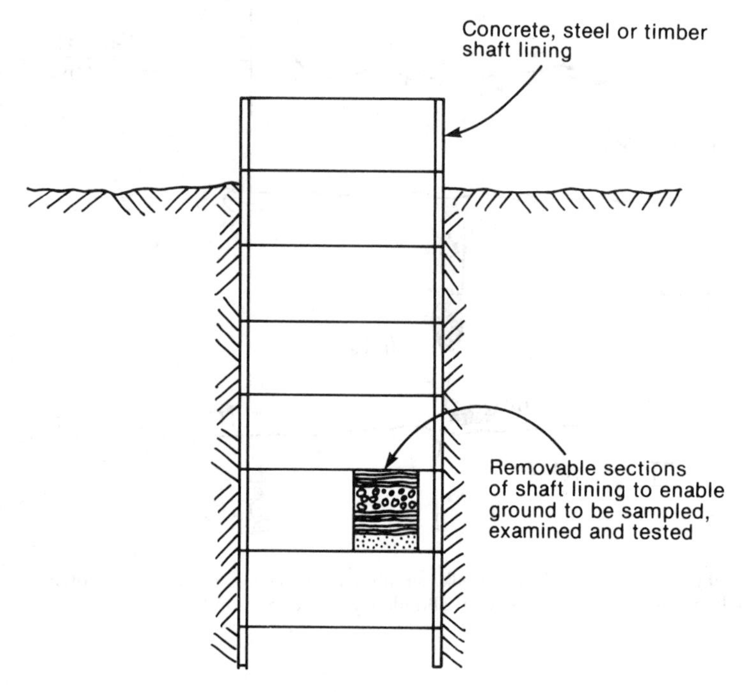

Fig. 9.7 Trial shaft in soil (diagrammatic).

from an engineering geologist's field notebook of a man-access borehole logging operation. The abbreviations vj and hj stand for vertical joint and horizontal joint, the symbol GII refers to the chalk weathering grade (see Table 5.7) and MH means medium hard chalk. About one-third of the borehole wall (120°) has been logged.

9.4 Shafts

During the site investigation for large or important underground structures trial shafts may be sunk and it will be the job of the engineering geologist to log them. If the trial shaft is in soil (Fig. 9.7) it will be supported by timber boarding or steel sheeting (BS 6031 : 1981) which will prevent the ground from being examined. Special provision will have to be made, therefore, for certain of the boards or sheets to be removable so that the engineering geologist may log the ground, carry out *in situ* tests and take samples. If the trial shaft is in rock the support may be of a kind (e.g. wire mesh and rock bolts) which allows almost continuous observations to be made, or there may be no support if the rock is sound. Access to the walls of a trial shaft may be by ladders with small landings at junctions of the ladders, but a far better arrangement

Fig. 9.8 Trial shaft with cable-supported travelling stage for examination.

from the logging point of view is to have a travelling stage supported by cables running up to winches at the surface as shown in Fig. 9.8. This enables the geologist with his equipment and lighting to be raised to the surface in increments of, say, 2 m at a time while he compiles his log. A surveyor's tape should be used to provide accurate depth measurement. Use of a travelling stage allows inspection of all parts of the shaft walls which is not possible with the fixed ladder arrangement. As for other exposures, photographs should be taken, and a flash gun will be needed for illumination. The methods for logging trial shafts are those described in Chapters 4 and 5; do not forget *in situ* strength tests can be made, and any ground-water encountered should be noted. If the geologist is present while the shaft is being sunk, it can, of course, be logged in a similar manner to logging a borehole (see Chapter 8).

9.5 Adits

In mining areas it may be possible to examine the geology by entering any adits that may be present. If the adit is for a working or recent mine there may be plans available which the engineering geologist should get hold of and use as a base map on which to record his observations. If no plans are available, the geologist should first chart the workings proposed to be examined, using a surveyor's tape for distances and a prismatic or Brunton compass for directions (see Sections 2.3 and 2.4.2). Be careful that any iron or steel in the workings do not affect the compass; this applies to other situations as well (see Section 9.6). The scale chosen will depend on the degree of detail required and the complexity of the geology and a quick reconnaissance should be made before deciding on the scale to be used. In mapping underground workings it is a convention to record the position of features at chest height, and this applies to geological details as well; however, if the geological features have low dips, their positions on the roof and floor of the adit will be considerably different to their positions shown on the map and this fact should be noted on the map for the benefit of any non-geologists who may later use it. If the adit is sloping it can still be mapped as though it were horizontal but the slope should be measured and recorded on the map, as should any positions of change in slope. In addition, the walls of the adit should be carefully examined and sketched in a series of vertical sections. Scanlines, photography, sampling, *in situ* testing and all the techniques previously described can be used where appropriate. Field notebooks, charts, camera and flash gun, etc. may all have to be protected from dripping water if the adit is wet. Safety helmets and cap lamps are, of course, essential but also take some powerful portable lanterns to illuminate the surfaces that are to be mapped and logged.

Before entering old adits the geologist should make sure that it is safe to do so. If the adit is unsupported be on the lookout for falls of rock on the floor which will indicate the roof or walls are unstable. If it is lined, examine the condition of the lining to see that it is still secure and doing its job. Particularly, you should suspect timber lining in wet conditions. An especial danger in old workings is the presence of timber staging. These are prone to rot or become insecure, and if covered with a layer or rock debris, their presence may be undetected until someone falls through them. In this situation a length of rope may be the means of saving life and anyone exploring old underground workings should be thus equipped. Never enter old workings alone and always inform a responsible person at the office when such entries are to be made.

For the site investigation of important underground works new adits are sometimes driven to gain access to geological formations which may not outcrop at the surface. Headings may be driven from the adit at a different levels of interest as shown in Fig. 9.9. All the remarks made about geotechnical observations in old adits apply equally to the examination of new adits except that in new ones the rock exposed will be fresh and not have suffered any deterioration which it might have done in old ones. The engineering geologist should be on the look out for, and note, any spalling or ravelling of the rock surface which takes place after excavation, and if he can, should observe the 'stand-up time' – that is the duration for which an unsupported length of the workings will stand unsupported. Rock liable to slaking can be identified by placing an air-dried specimen in water and observing if it breaks down into fragments.

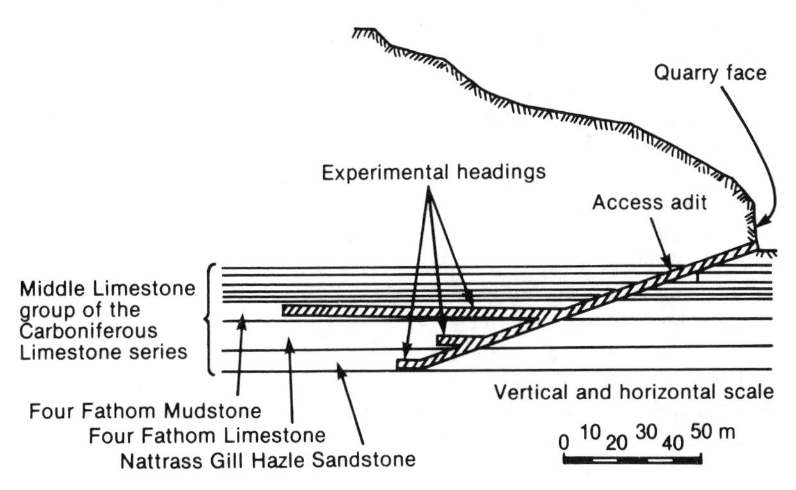

Fig. 9.9 Adit and headings driven to explore underground strata.

9.6 Tunnels

There are two elements to tunnel logging – the compiling of a continuous record of the geology of the drive and the recording of geotechnical properties, usually done by making measurements or observations at specific representative locations. Observations can be made during excavation at the tunnel face, and after excavation along the tunnel walls. The method of tunnel construction will greatly affect the degree of logging possible. In a soft ground tunnel being driven with a full-face tunnelling machine and being lined with pre-cast concrete segments the scope for logging will be almost nil but in a hard rock tunnel being driven by drill-and-blast with little support there may be almost continuously uninterrupted exposure enabling very complete logs to be compiled.

The first task is to make a record of the geology, either in the form of face sections or of tunnel wall sections. The stratigraphy, lithology and structure should all be systematically noted. A scale should always be provided on the drawings and for this purpose convenient features such as steel rib arches, often set at 1 m spacing, should be indicated. These sections can be drawn in the field notebook but it is becoming increasingly common practice to use A4-sized geological log sheets instead. These allow the drawing of the face or wall section to be made on a prepared gridded outline and the lithological and other details of the rock types to be recorded by filling in boxes instead of by extensive note taking (Figs 9.10, 9.11 and 9.12). Figs 9.10 and 9.11 show logsheets for rock and Fig. 9.12 a logsheet for a mixed face, in this case mainly soil. One advantage of this system is that the blank geological log sheet acts itself as a reminder of what observations must be made. The log sheet blanks should be drawn up to suit the geology of the particular tunnel being logged and the particular geotechnical observations that are required. If log sheets are used they should be securely held in a strong clip board fitted with a waterproof cover, and an index should be compiled of the individual sheets for collation of them later on back in the office. As with all other logging, photographic records should supplement your drawings – remember that a flash gun will be needed for illumination.

The geotechnical properties that are measured are the intact rock strength and the discontinuities. Strength can be estimated *in situ* using the Schmidt hammer (Fig. 9.13) or on samples taken from the tunnel using the point load strength or cone indenter apparatus (see Section 6.1). Vertical or horizontal scanlines can be put up to measure spacing and orientation of discontinuities as described in Section 5.2. The selection of positions for making *in situ* measurements and taking samples should be done systematically. One satisfactory way of doing this is to divide the tunnel length into zones, within each one of which the geological conditions are broadly uniform, and then make sure

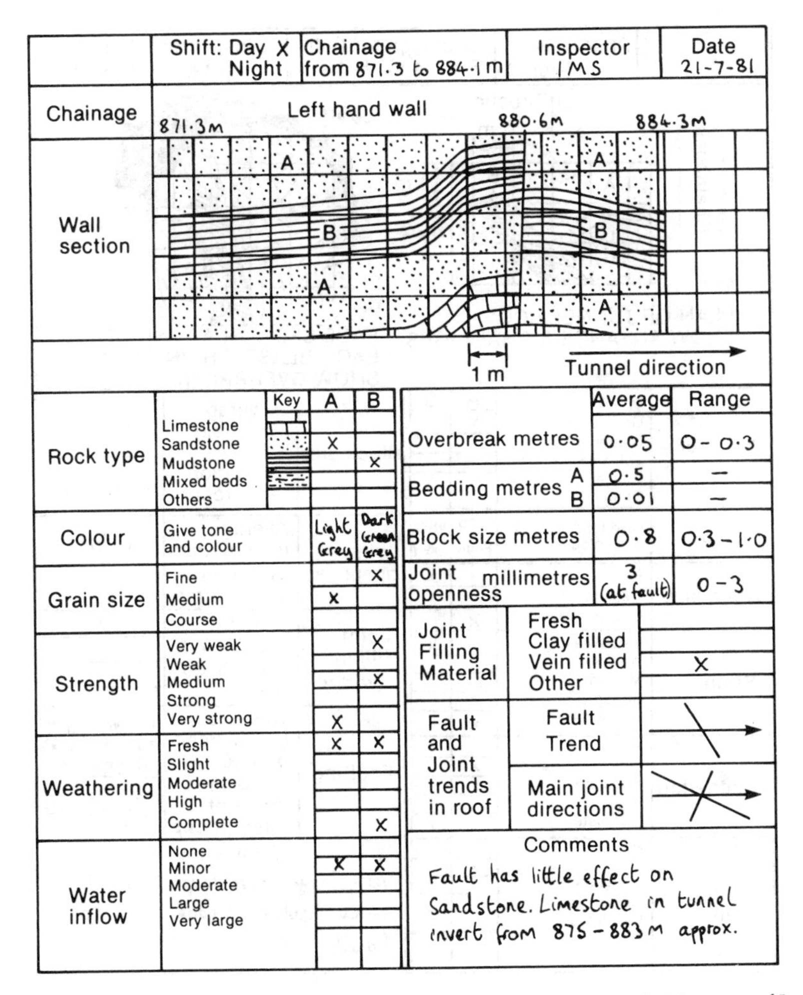

Fig. 9.10 Logsheet for recording geology of tunnel wall section in rock. (Courtesy of I. McFeat-Smith, Charles Haswell and Partners.)

that there are sufficient sampling points situated in each zone to characterize each rock type present and that the points chosen are representative of the rocks. In tunnels lined with wire mesh, lagging, etc. it is often

	Shift: Day Night X	Chainage for Pull From 312m to 314·2m	Inspector IMS	Date 30-11-79

		A	B
Rock type	Granite	X	
	Granodiorite		
	Volcanic		
	Gouge		X
	Other		
Colour	Give tone and colour	MED. PINK	DARK GREY
Grain size	Fine		X
	Medium		
	Course	X	
Strength	Very Weak		X
	Weak		
	Medium	X	
	Strong	X	
	Very strong	X	
Weathering	Fresh		
	Slight	X	
	Moderate	X	
	High		
	Complete		X
Water inflow	None		
	Minor	X	
	Moderate		
	Large		
	Very large		X

Block size	Average (metres)	0·2
	Variation (metres)	0·05 – 0·4
Joint openness	Average (millimetres)	TIGHT
	Variation (millimetres)	
Joint filling material	Fresh	X
	Clay filled	
	Vein filled	
	Other	
Faulting	Observed (show trend on plan)	1m wide gouge zone with blocks of granite
	Possible (type of disruption)	Increased jointing & water in A
Comments	Very large overbreak in right hand shoulder & roof due to fault.	

Fig. 9.11 Logsheet for recording geology of tunnel face section in rock. (Courtesy of I. McFeat-Smith, Charles Haswell and Partners.)

convenient to choose one of the refuges cut out of the tunnel wall, if they are provided, as a sampling point because of the good access to the rock.

In addition to strength and discontinuity measurements there are a number of important features that should be specifically looked for and re-

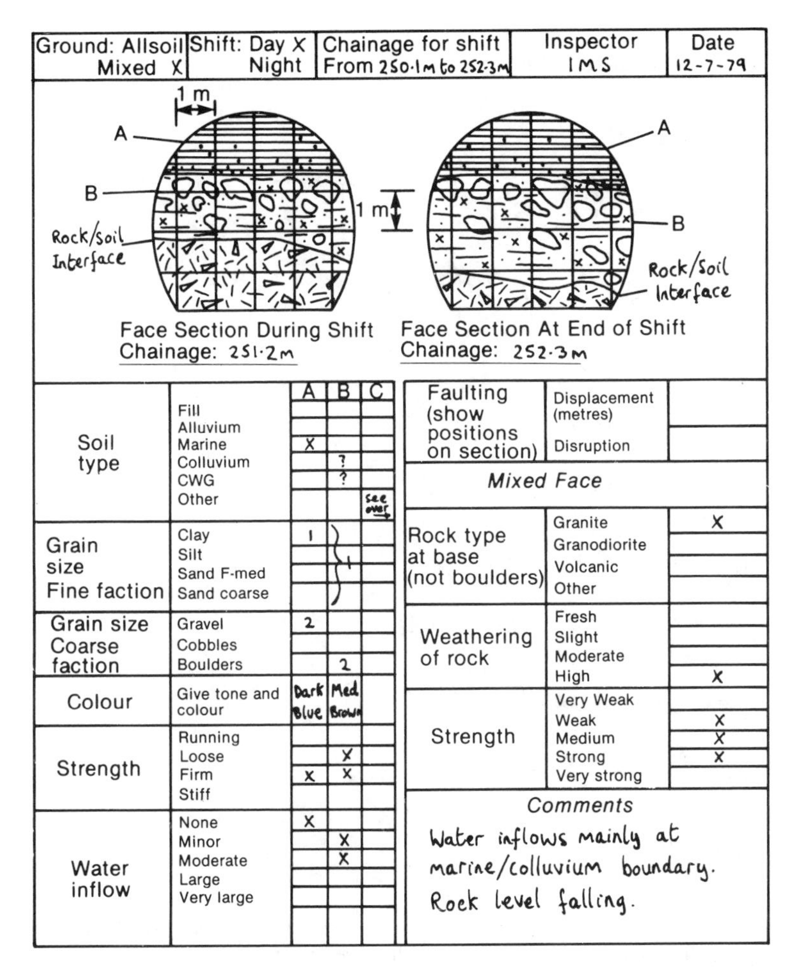

Ground: Allsoil / Mixed X	Shift: Day X / Night	Chainage for shift From 250·1m to 252·3m	Inspector IMS	Date 12-7-79

		A	B	C
Soil type	Fill			
	Alluvium			
	Marine	X		
	Colluvium		?	
	CWG		?	
	Other			see over
Grain size Fine faction	Clay	1		
	Silt		1	
	Sand F-med			
	Sand coarse			
Grain size Coarse faction	Gravel	2		
	Cobbles			
	Boulders		2	
Colour	Give tone and colour	Dark Blue	Med Brown	
Strength	Running			
	Loose		X	
	Firm	X	X	
	Stiff			
Water inflow	None	X		
	Minor		X	
	Moderate		X	
	Large			
	Very large			

Faulting (show positions on section)	Displacement (metres)	
	Disruption	
Mixed Face		
Rock type at base (not boulders)	Granite	X
	Granodiorite	
	Volcanic	
	Other	
Weathering of rock	Fresh	
	Slight	
	Moderate	
	High	X
Strength	Very Weak	
	Weak	X
	Medium	X
	Strong	X
	Very strong	

Comments

Water inflows mainly at marine/colluvium boundary. Rock level falling.

Fig. 9.12 Logsheet for recording geology of tunnel face section in mixed ground, mainly soil. (Courtesy of I. McFeat-Smith, Charles Haswell and Partners.)

corded. These are faults, zones of broken rock, the presence of gouge or bands of soft material, overbreak (Fig. 9.14), ravelling, spalling or slabbing from the tunnel walls, instability in the roof, and any irregularity in the tunnel profile that may have a geological cause. Also, note any positions

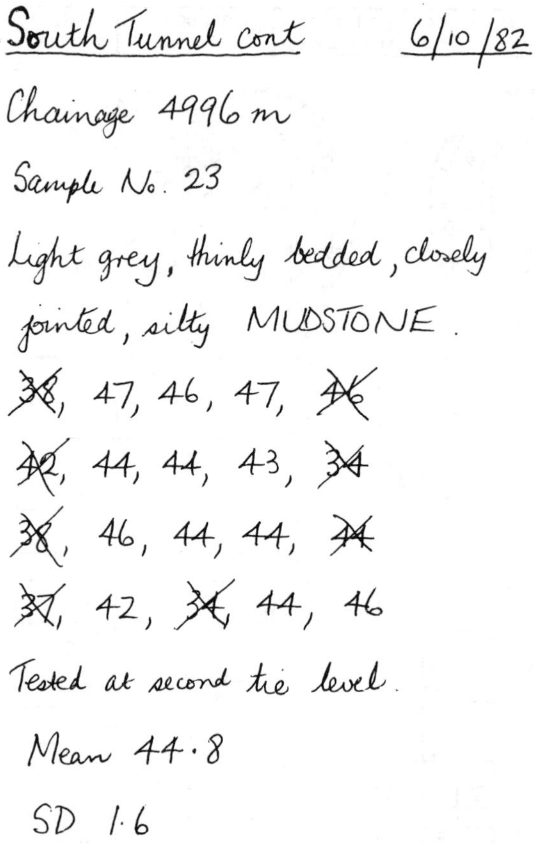

South Tunnel cont   6/10/82

Chainage 4996 m

Sample No. 23

Light grey, thinly bedded, closely jointed, silty MUDSTONE.

~~38~~, 47, 46, 47, ~~46~~

~~42~~, 44, 44, 43, ~~34~~

~~38~~, 46, 44, 44, ~~44~~

~~37~~, 42, ~~34~~, 44, 46

Tested at second tie level.

Mean 44·8

SD 1·6

Fig. 9.13 Page from a field notebook. The geologist has recorded Schmidt hammer readings: five values at each of four spots. The lowest two from each set of five have been rejected and the mean and standard deviation calculated from the rest. (Low readings with the Schmidt hammer are often due to unsatisfactory test surface conditions.)

where water enters the tunnel and record the intensity of inflow with some suitable semi-quantitative scale such as: spouts, rain, drippers, seepage, damp. Alternatively, using a bucket and a watch, the inflow can be quantified. In areas of difficulty with ground conditions during construc-

Fig. 9.14 Overbreak in a tunnel wall. A block of rock, defined by the intersection of two joints and one bedding plane parting with the tunnel wall, has fallen out leaving a cavity.

tion special attention to detailed logging and observations should be given with the aim of trying to establish the cause of the difficulty. The performance of the primary support should be observed and any shortcomings or distress noted. When using a compass in a tunnel remember that it will be affected by any nearby metal. As a rough guide, Yow (1982) says that compass readings should not be taken within 30 cm of wire mesh, 45 cm of a rock bolt, 60 cm of tunnelling equipment such as a drilling jumbo and nowhere beneath a steel arch rib.

When logging a tunnel, comply with BS 614 : 1990 and follow all the advice given previously about safety, particularly the remarks made on working in adits given in Section 9.5. Also remember that in a working tunnel there will be additional hazards from plant in use such as tunnelling machines, lining erection machinery, drilling jumbos, moving conveyor belts, locomotives and muck wagons, crossing points, electricity, compressed air, hydraulic systems, etc. and follow any special directives given by the contractor. Some tunnels through Coal Measures strata are driven under coal mining safety regulations; the engineering geologist needs to note that the effect of these will prevent the use underground of flash photography and any equipment made from light alloy (see note on Schmidt hammer in Section 6.1).

9.6.1 *Benches*

Some excavations, e.g. open pit mines and excavations for dam foundations, are made in a series of benches, each of which consists of an exposed horizontal surface and exposed nearly vertical face. Because of their large size and complexity the geotechnical logging of these is a specialized subject outside the scope of this book, but detailed advice on appropriate techniques is given by Knill and Jones (1965). However, a few general remarks will be made on aspects where methods discussed in earlier sections of the book are relevant. If the horizontal surface of the bench is free from debris or is not being used as a haul road it can be mapped, and the face of the bench can be logged using the same methods as for a trial pit face (see Section 9.1). If the bench face is very high it will be possible to log only the lower accessible part but the whole face can, and should, be recorded photographically as described in Section 7.2.1.

The geological mapping of the bench surfaces can be done using plans of the site as base maps, and the logging of the faces can be recorded on elevation drawings if these exist. If not, the engineering geologist will have to prepare a set of gridded sheets on which scale drawings of the bench faces can be made. Remember to identify each one clearly so that they can be collated later in the office. Also record the compass bearing of the face being logged so that it can be orientated correctly. The height of bench faces can be measured directly if low, or indirectly if high using the Abney level as described in Section 2.2.

9.7 Construction records

The recording of geological and geotechnical conditions during construction of civil engineering works is done for the following reasons: (1) to confirm or supplement the site investigation predictions, (2) to provide a basis for contractual payment, especially for ascertaining 'unexpected' ground conditions, (3) to determine the effects of ground conditions on excavation and construction and (4) to provide a permanent record of the as-constructed ground conditions before the excavations, tunnel walls, etc. are permanently hidden from view by the foundations, linings, etc. of the built structure itself. The kind of logging will therefore be of the general kind supplemented by the specific as needs dictate. All the remarks made earlier in this chapter apply equally to the compilation of construction records, the only difference being perhaps one of emphasis, the engineering geologist being particularly alert to record exposures and evidence that may be inaccessible evermore.

Examples of construction record *drawings* are shown in Figs 9.15 and 9.16. These are for an underground railway, and a continuous record of joints, bedding planes, faults, roof instability and ground-water flow was made as construction was carried out. The colour and general condition of the rock was assessed and any bands or layers of sand, mica or clay were noted. The length and inclination of joints and bedding planes were measured together with width of parting, and a note made of whether they were open, filled or closed. Similarly, all fault characteristics were determined and any associated shattered zones noted. An assessment of the severity of overbreak in crown or sides was made and its possible cause recorded. All the information was entered on the longitudinal sections of the tunnel as shown in Fig. 9.15. In addition, face sections as shown in Fig. 9.16 were drawn for the tunnel face every 30–40 m. These gave a detailed description of the geology of the face, the condition of the rock and the results of Schmidt hammer tests.

Examples of construction record *photographs* are shown in Figs 9.17 and 9.18. These are of a highway rock cutting, and the vertical pattern of fluting on the rock face is produced by the boreholes drilled for pre-split blasting. The scale is given by various items of plant in the photographs. Figures 9.17 and 9.18 are also examples of a three-shot panorama photograph, the taking of which was described in Section 7.2.1. Figure 9.17 shows an overall view of the rock cutting with the scale changing markedly towards the right of the picture, whilst Fig. 9.18 shows a section of the rock cutting in more detail, and with less change in scale because the camera angles do not diverge as much as in Fig. 9.17.

Figures 9.15 to 9.18 will suggest to you how you can make both construction drawings and construction photographs as construction records for your own projects.

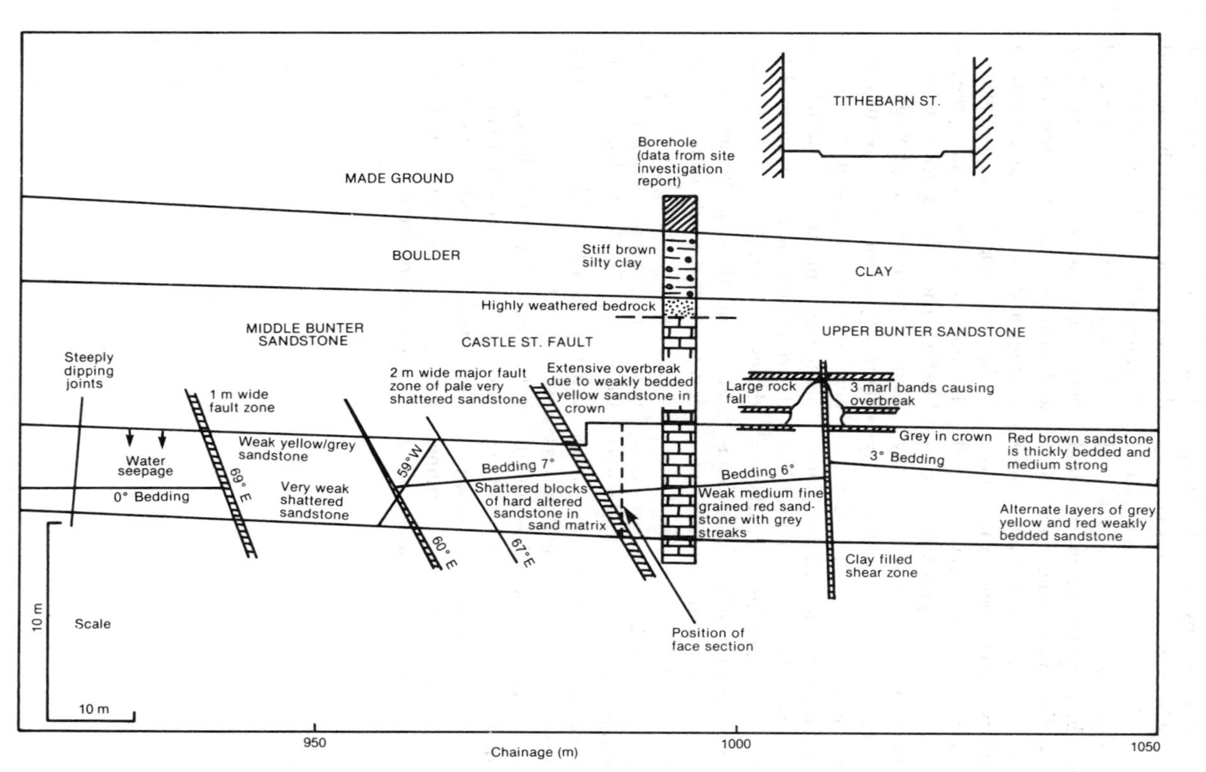

Fig. 9.15 Longitudinal section of part of an underground railway tunnel showing ground information obtained during construction. A site investigation borehole column is also included.

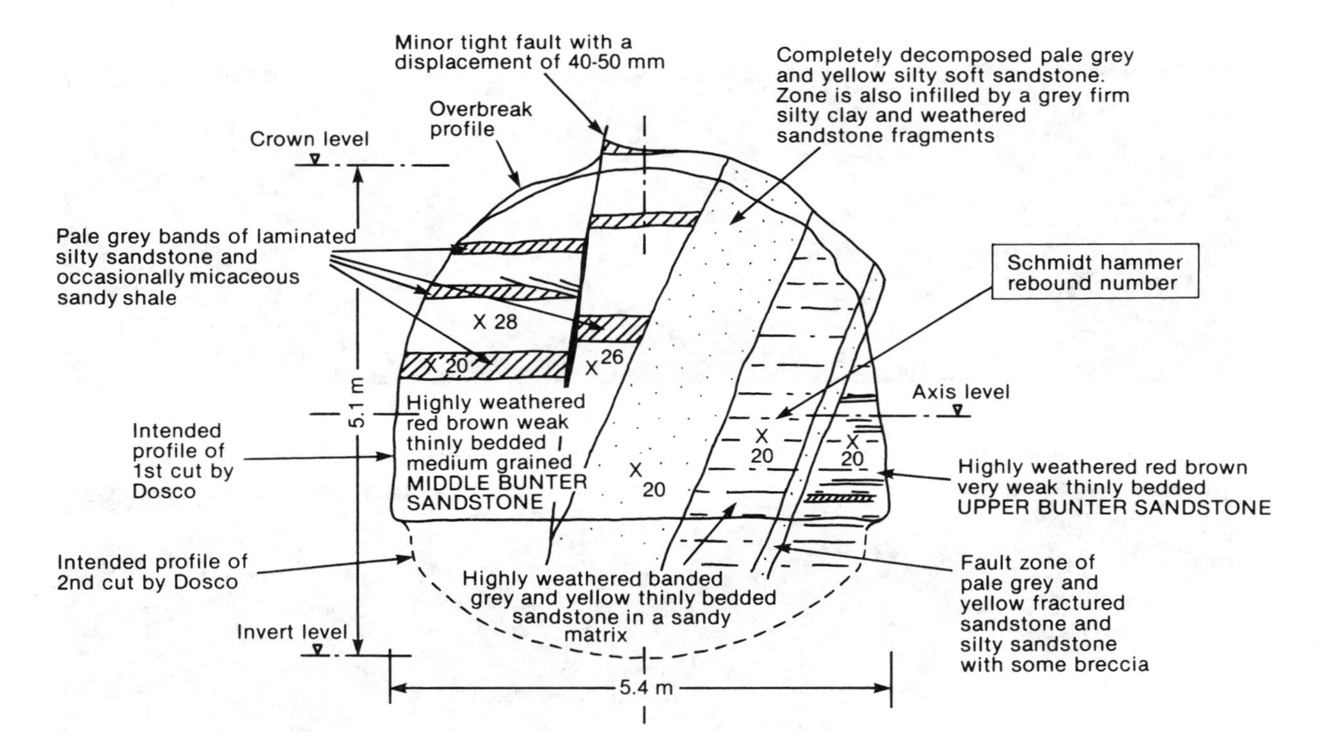

Fig. 9.16 Face section prepared during tunnel construction showing geological information and Schmidt hammer test results. The position of the face section is indicated on the longitudinal section in Fig. 9.15.

Fig. 9.17 Construction record photograph of highway rock cutting made using three-shot panorama technique. Note change in scale.

Fig. 9.18 Pre-split blasting face of highway rock cutting in high-grade metamorphic rock, mainly quartzite.

Bibliography

This bibliography consists of three lists of books and papers. List A are those that you should obtain for your own personal use. List B are those that should be readily available for consultation in your organization; the British Standard Codes of Practice and the United States Standard OSHA 2207 *must* be available. List C are others that will be useful, including those that are referred to in the text of this book or are acknowledged as the source of illustrations or tables. Remember also the other handbooks in this series listed at the front of this book, which could be added to List A or B.

List A

BRITISH STANDARDS INSTITUTION (1981) *Code of practice for site investigations*. British Standard BS 5930:1981. London, British Standards Institution.

DUMBLETON, M. J. & WEST, G. (1976) 'Preliminary sources of information for site investigations in Britain.' *TRRL Report* LR 403 (revised edition). Crowthorne, Transport and Road Research Laboratory.

GEOLOGICAL SOCIETY ENGINEERING GROUP WORKING PARTY (1970) 'The logging of rock cores for engineering purposes.' *Quarterly Journal of Engineering Geology*, 3(1), 1–24.

GEOLOGICAL SOCIETY ENGINEERING GROUP WORKING PARTY (1972) 'The preparation of maps and plans in terms of engineering geology.' *Quarterly Journal of Engineering Geology*', 5(4), 293–382.

GEOLOGICAL SOCIETY ENGINEERING GROUP WORKING PARTY (1977) 'The description of rock masses for engineering purposes.' *Quarterly Journal of Engineering Geology*, 10(4), 355–88.

WELTMAN, A. J. & HEAD, J. M. (1983) 'Site investigation manual.' *CIRIA Special Publication* 25. London, Construction Industry Research and Information Association.

List B

AMERICAN SOCIETY FOR TESTING AND MATERIALS (1988) *Annual Book of ASTM Standards.* Section 4: Construction. Volume 04.08. Philadelphia, American Society for Testing and Materials.

BRITISH STANDARDS INSTITUTION (1972) *Code of practice for foundations.* Code of Practice CP 2004:1972. London, British Standards Institution.

BRITISH STANDARDS INSTITUTION (1975) *Methods of test for soils for civil engineering purposes.* British Standard BS 1377:1975. London, British Standards Institution.

BRITISH STANDARDS INSTITUTION (1978) *Code of practice for safety precautions in the construction of large diameter boreholes for piling and other purposes.* British Standard BS 5573:1978. London, British Standards Institution.

BRITISH STANDARDS INSTITUTION (1981) *Code of practice for earthworks.* British Standard BS 6031:1981. London, British Standards Institution.

BRITISH STANDARDS INSTITUTION (1990) *Safety in tunnelling in the construction industry.* British Standard BS 614:1990. London, British Standards Institution.

BROWN, E. T. (1981) *Rock Characterization Testing and Monitoring: ISRM Suggested Methods.* Oxford, Pergamon Press.

DUMBLETON, M. J. & WEST, G. (1970) 'Air-photograph interpretation for road engineers in Britain.' *RRL Report* LR 369. Crowthorne, Road Research Laboratory.

DUMBLETON, M. J. & WEST, G. (1974) 'Guidance on planning, directing and reporting site investigations.' *TRRL Report* LR 625. Crowthorne, Road Research Laboratory.

IRVINE, D. J. & SMITH, R. J. H. (1983) 'Trenching practice.' *CIRIA Report* 97. London, Construction Industry Research and Information Association.

LAHEE, F. H. (1961) *Field Geology*, 6th edn. London, McGraw-Hill.

UNITED STATES DEPARTMENT OF LABOR (1983) *Occupational Safety and Health Administration Standard OSHA 2207.* Washington DC, US Government Printing Office.

List C

AMERICAN SOCIETY FOR TESTING AND MATERIALS (1983) *Annual Book of ASTM Standards.* Section 4: Construction. Volume 04.08. Philadelphia, American Society for Testing and Materials.

BATES, D. E. B. & KIRKALDY, J. F. (1976) *Field Geology in Colour.* Poole, Blandford Press.

Bibliography

BROCH, F. & FRANKLIN, J. A. (1972) 'The point-load strength test.' *International Journal of Rock Mechanics and Mining Sciences*, 9, 669–97.

CARTER, P. G. & MILLS, D. A C. (1976) 'Engineering geological investigations for the Kielder tunnels.' *Quarterly Journal of Engineering Geology*, 9(2), 125–41.

COMPTON, R. R. (1962) *Manual of Field Geology*, Chapter 2. London, John Wiley.

DAVIES, T. P., CARTER, P. G., MILLS, D. A. C. & WEST, G. (1981) 'Kielder aqueduct tunnels – predicted and actual geology.' *TRRL Report* SR 676. Crowthorne, Transport and Road Research Laboratory.

HATHEWAY, A. W. (1982) 'Trench, shaft and tunnel mapping.' *Bulletin of the Association of Engineering Geologists*, 19(2), 173–80.

HOEK, E. & BRAY, J. W. (1977) *Rock Slope Engineering*, Chapter 4. London, The Institution of Mining and Metallurgy.

JOYCE, M. D. (1982) *Site Investigation Practice*, Chapter 5. London, E. & F. N. Spon.

KNILL, J. L. & JONES, K. S. (1965) 'The recording and interpretation of geological conditions in the foundations of the Roseires, Kariba, and Latiyan Dams. *Géotechnique*, 15(1), 94–124.

LEROY, L. W., LEROY, D. O. & RAESE, J. W. (1977) *Subsurface Geology: Petroleum, Mining, Construction*, 4th edn. Golden, Colorado, Colorado School of Mines.

MCFEAT-SMITH, I. (1982) 'Logging tunnel geology.' *Tunnels and Tunnelling*, 14(4), 20–25.

MOSELEY, F. (1981) *Methods in Field Geology*, Oxford, W. H. Freeman.

NATIONAL COAL BOARD (1972) NCB cone indenter. *MRDE Handbook* No. 5. Bretby, Mining Research and Development Establishment.

ROAD RESEARCH LABORATORY (1952) *Soil Mechanics for Road Engineers*. London, HM Stationery Office, pp. 369–73.

THOMPSON, M. M. (1987) *Maps for America*, 3rd edn. Reston, Virginia, US Geological Survey.

UNITED STATES DEPARTMENT OF COMMERCE (1960) *The Identification of Rock Types* (revised edition). Washington, DC, US Government Printing Office.

WAKELING, T. R. M. (1969) 'A comparison of the results of standard site investigation methods against the results of a detailed geotechnical investigation in the Middle Chalk at Mundford, Norfolk.' *Proceedings of the Conference on In Situ Investigations in Soils and Rocks*. London, British Geotechnical Society, pp. 17–22.

WARD, W. H., COATS, D. J. & TEDD, P. (1976) 'Performance of tunnel support systems in the Four Fathom Mudstone.' *Tunnelling 76*. London, The Institution of Mining and Metallurgy, pp. 329–40.

WEEKS, A. G. (1969) 'The stability of natural slopes in south-east England as affected by periglacial activity.' *Quarterly Journal of Engineering Geology*, 2(1), 49–61.

WILLIAMS, J. C. C. (1969) *Simple Photogrammetry*. London, Academic Press.

WILLIAMSON, D. A. & KUHN, R. (1988) 'The Unified Rock Classification system.' In *Rock Classification Systems for Engineering Purposes*, STP 984. Philadelphia, American Society for Testing and Materials, pp. 7–16.

WRIGHT, J. (1982) *Ground and Air Survey for Field Scientists*. Oxford, Clarendon Press.,

YOW, J. L. (1982) 'Sources of compass error in tunnel mapping.' *Bulletin of the Association of Engineering Geologists*, 19(2), 133–40.

Availability

The following addresses will be useful in obtaining some of the publications listed:

Great Britain

British Standards Institution, Linford Wood, Milton Keynes MK14 6LE.

Construction Industry Research and Information Association, 6 Storey's Gate, London SW1P 3AU.

The Geological Society Publishing House, Unit 7, Brassmill Enterprise Centre, Brassmill Lane, Bath, BA1 3JN.

The Institution of Mining and Metallurgy, 44 Portland Place, London W1N 4BR.

Transport and Road Research Laboratory, Old Wokingham Road, Crowthorne, Berkshire RG11 6AU.

United States

American Society for Testing and Materials, 1916 Race Street, Philadelphia, PA 19103.

Association of Engineering Geologists, 1041 New Hampshire Street, Lawrence, Kansas, KS 66044.

US Government Printing Office, Washington, DC 20402.

Index